Lecture Notes in Mathematics

Edited by A. Dold and B. Eckmann

1169

W. A. Light
E. W. Cheney

T0224601

Approximation Theory
in Tensor Product Spaces

Springer-Verlag
Berlin Heidelberg New York Tokyo

Authors

William Allan Light
Mathematics Department, University of Lancaster
Bailrigg, Lancaster LA1 4YL, England

Elliott Ward Cheney
Mathematics Department, University of Texas
Austin, Texas 78712,USA

Mathematics Subject Classification (1980): Primary: 41A63, 41A65
Secondary: 41-02, 41A30, 41A45, 41A50

ISBN 3-540-16057-4 Springer-Verlag Berlin Heidelberg New York Tokyo
ISBN 0-387-16057-4 Springer-Verlag New York Heidelberg Berlin Tokyo

Printing and binding: Beltz Offsetdruck, Hemsbach/Bergstr.
2146/3140-543210

DEDICATION

This work is dedicated to the memory of

Robert Schatten

(1911 - 1977)

who did much of the pioneering work in

the theory of tensor products of Banach spaces.

PREFACE

In the past two decades, a new branch of approximation theory has emerged; it concerns the approximation of multivariate functions by combinations of univariate ones. The setting for these approximation problems is often a Banach space which is the tensor product of two or more simpler spaces. Approximations are usually sought in subspaces which are themselves tensor products. While these are infinite dimensional, they may share some of the characteristics of finite-dimensional subspaces. The usual questions from classical approximation theory can be posed for these approximating subspaces, such as (i) Do best approximations exist? (ii) Are best approximations unique? (iii) How are best approximations characterized? (iv) What algorithms can be devised for computing best approximations? (v) Do there exist simple procedures which provide "good" approximations, in contrast to "best" approximations? (vi) What are the projections of least norm on these subspaces? and (vii) what are the projection constants of these subspaces?

This volume surveys only a part of this growing field of research. Its purpose is twofold: first, to provide a coherent account of some recent results; and second, to give an exposition of the subject for those not already familiar with it. We cater for the needs of this latter category of reader by adopting a deliberately slow pace and by including virtually all details in the proofs. We hope that the book will be useful to students of approximation theory in courses and seminars.

Expert readers may wish to omit a reading of the first chapter, which gives an introduction to the tensor product theory of Banach spaces. The material on approximation theory occupies the next eight chapters. Results needed in proofs but perhaps not familiar to every reader are collected in two appendices (Chapters 10 and 11). Finally, there are historical notes and a large collection of references, some of which are only peripheral to our theme.

Notation and conventions are standard throughout, and we often do not stop to define notation which we expect to be familiar. However, a table of notation has been placed just before the index.

We are glad to be able to thank a number of colleagues for pleasant collaboration over the years on matters relating to these notes: Carlo Franchetti (Florence), Manfred von Golitschek (Würzburg), Julie Halton (Lancaster), Sue Holland (Lancaster), John Respess (Austin), and Lin Sulley (Lancaster and Ipswich).

During the preparation of the manuscript, the second author was supported by grants from the University of Texas and the Science and Engineering Research Council of Great Britain. For these grants, and for the hospitality of the University of Lancaster, he is deeply grateful.

We are very much indebted to Ms. Jan Duffy of the University of Texas Mathematics Department, who undertook the arduous task of rendering our manuscript into a computer file for processing by the TEX typesetting system. The pleasing appearance which (we think) the book possesses is due entirely to the skill and good judgement of Ms. Duffy.

W.A. Light   E.W. Cheney

Lancaster, July 1985

CONTENTS

CHAPTER 1

AN INTRODUCTION TO TENSOR PRODUCT SPACES

The purpose of this chapter is to introduce some of the basic theory of the tensor product of two Banach spaces. All of this material can be found in other sources, but the treatment here is particularly designed to meet the needs of the subsequent chapters. As mentioned in the preface, we have tried to expound the subject in a way that leaves very few arguments for the reader to supply.

There are two sources of information on this topic which we should mention at the outset. One is Schatten's monograph [154], which gives a very careful treatment of the foundations of the subject. The other is the survey of vector measures by Diestel and Uhl [55]. Chapter 8 of [55] provides a brief introduction to tensor products of Banach spaces and covers many recent results in this area.

Let X and Y be Banach spaces, and denote their duals by X^* and Y^*, respectively. We shall construct formal expressions $\sum_{i=1}^{n} x_i \otimes y_i$ where $x_i \in X$, $y_i \in Y$ and $n \in \mathbb{N}$. We will regard such an expression as defining an operator $A : X^* \to Y$, given by

$$A\phi = \sum_{i=1}^{n} \phi(x_i) y_i \qquad (\phi \in X^*).$$

Amongst all these formal expressions we introduce the relation

$$\sum_{i=1}^{n} x_i \otimes y_i \sim \sum_{i=1}^{m} a_i \otimes b_i$$

if both expressions define the same operator from X^* to Y. This is clearly an equivalence relation on the set of all such formal expressions. We shall henceforward be interested only in the equivalence classes of this relation, and will denote the set of all such equivalence classes by $X \otimes Y$. We shall abuse notation in the usual way by referring to the expression $\sum_{i=1}^{n} x_i \otimes y_i$ as a member of $X \otimes Y$ when we intend to refer to the equivalence class of expressions containing $\sum_{i=1}^{n} x_i \otimes y_i$. For any $\alpha \in \mathbb{R}$ we define multiples of (the equivalence class of expressions containing) $\sum_{i=1}^{n} x_i \otimes y_i$ by (the equivalence class of expressions containing) $\sum_{i=1}^{n} \alpha x_i \otimes y_i$. Similarly, we define addition by

$$\sum_{i=1}^{n} x_i \otimes y_i + \sum_{i=n+1}^{m} x_i \otimes y_i = \sum_{i=1}^{m} x_i \otimes y_i.$$

All algebraic identities in $X \otimes Y$ are based upon the interpretation of expressions as linear operators. Thus one easily verifies such identities as

$$x \otimes (u + v) = x \otimes u + x \otimes v$$

$$\alpha x \otimes y = x \otimes \alpha y$$

$$x \otimes 0 = 0 \otimes 0.$$

Two observations are perhaps helpful at this juncture. Firstly, the symbol "+" is being used only as a separator in our formal expression $\sum_{i=1}^{n} x_i \otimes y_i$ and in our definition of the addition of two such expressions. Secondly, the construction thus far makes no use of the topological structure of X or Y. Thus we could have begun with linear spaces and algebraic duals rather than Banach spaces. If we begin with Banach spaces and identify $\sum_{i=1}^{n} x_i \otimes y_i$ with an operator A mapping the algebraic dual of X into Y then the equivalence classes that make up $X \otimes Y$ remain the same.

It is clear that scalar multiplication of $\sum_{i=1}^{n} x_i \otimes y_i$ by α is equivalent to multiplying the associated operator A by α, and that the addition of this expression to $\sum_{j=1}^{m} a_j \otimes b_j$ is equivalent to adding A to the operator B associated with this last expression. With these definitions $X \otimes Y$ forms a linear space, called the **algebraic tensor product**.

1.1 LEMMA. *Every expression $\sum_{i=1}^{n} x_i \otimes y_i$ is equivalent to either $0 \otimes 0$ or to an expression $\sum_{i=1}^{m} a_i \otimes b_i$ where $\{a_1, \ldots, a_m\}$ and $\{b_1, \ldots, b_m\}$ are linearly independent sets.*

PROOF. Suppose that for instance $x_n = \sum_{j=1}^{n-1} \alpha_j x_j$. Then $\sum_{i=1}^{n} x_i \otimes y_i$ defines the operator $A : X^* \to Y$ where, for $\phi \in X^*$,

$$
\begin{aligned}
A\phi = \sum_{i=1}^{n} \phi(x_i)y_i &= \sum_{i=1}^{n-1} \phi(x_i)y_i + \phi(x_n)y_n \\
&= \sum_{i=1}^{n-1} \phi(x_i)y_i + \phi\left(\sum_{j=1}^{n-1} \alpha_j x_j\right)y_n \\
&= \sum_{i=1}^{n-1} \phi(x_i)y_i + \sum_{j=1}^{n-1} \alpha_j \phi(x_j)y_n \\
&= \sum_{i=1}^{n-1} \phi(x_i)(y_i + \alpha_i y_n).
\end{aligned}
$$

Hence $\sum_{i=1}^{n} x_i \otimes y_i$ has the representation $\sum_{i=1}^{n-1} a_i \otimes b_i$ where $a_i = x_i$ and $b_i = y_i + \alpha_i y_n$. We may repeat this process until we arrive at either a representation in which $\{a_1, \ldots, a_m\}$ and $\{b_1, \ldots, b_m\}$ are linearly independent or one of the representations $x \otimes 0$ or $0 \otimes y$, which are both equivalent to $0 \otimes 0$. ∎

2

The space $X \otimes Y$ is generated by elements of the form $x \otimes y$, which are called **dyads**. This observation is often used to simplify linear arguments.

From the preceding discussions it is clear that $X \otimes Y$ may be regarded as a subspace of the space of continuous linear operators of finite rank from X^* into Y. Usually this subspace is a proper one. However, in the case when X is reflexive every continuous finite rank linear operator from X^* into Y can be identified with an expression $\sum_{i=1}^{n} x_i \otimes y_i$ as follows. Suppose A is such an operator with range determined by linearly independent $b_1, \ldots, b_n$. Then for $\phi \in X^*$,

$$A\phi = \sum_{i=1}^{n} \alpha_i(\phi) b_i \quad \text{where} \quad \alpha_i(\phi) \in \mathbf{R}.$$

Standard arguments show that the α_i are in X^{**} and hence in X, whence $\alpha_i(\phi) = \phi(a_i)$ for suitable $a_i \in X$, $1 \le i \le n$. Hence,

$$A\phi = \sum_{i=1}^{n} \phi(a_i) b_i$$

and A is associated with $\sum_{i=1}^{n} a_i \otimes b_i$.

It is possible to construct various norms on $X \otimes Y$ using the norms in X and Y. The most obvious way to introduce a norm which is independent of the representation of the equivalence class is to assign to $\sum_{i=1}^{n} x_i \otimes y_i$ the norm it receives when regarded as an operator from X^* to Y; *viz.*,

$$\lambda\left(\sum_{i=1}^{n} x_i \otimes y_i\right) = \sup\left\{\left\|\sum_{i=1}^{n} \phi(x_i) y_i\right\| : \phi \in X^*, \; \|\phi\| = 1\right\}.$$

Notice that for a dyad $x \otimes y$ we have

$$\lambda(x \otimes y) = \sup\{\|\phi(x)y\| : \phi \in X^*, \; \|\phi\| = 1\}$$
$$= \sup\{|\phi(x)| \, \|y\| : \phi \in X^*, \; \|\phi\| = 1\}$$
$$= \|x\| \, \|y\|.$$

Such a norm on $X \otimes Y$, for which the norm of a dyad equals the product of the norms of its two components, is termed a **crossnorm**. Given any two Banach spaces X and Y there is a rich supply of crossnorms on $X \otimes Y$.

We can form a second linear space from X and Y by considering $X^* \otimes Y^*$. This consists of all expressions

$$\sum_{i=1}^{n} \phi_i \otimes \psi_i \quad \text{where} \quad \phi_i \in X^* \quad \text{and} \quad \psi_i \in Y^*.$$

3

In addition to other interpretations such an expression may be considered as a linear form on $X \otimes Y$ by defining

$$\left(\sum_{i=1}^{n} \phi_i \otimes \psi_i \right) \left(\sum_{j=1}^{m} x_j \otimes y_j \right) = \sum_{i=1}^{n} \sum_{j=1}^{m} \phi_i(x_j)\psi_i(y_j).$$

One verifies easily that this definition is *proper*; that is, it is invariant over the equivalence classes. Note that it would have sufficed to give the above definition for a pair of dyads:

$$(\phi \otimes \psi)(x \otimes y) = \phi(x)\psi(y).$$

1.2 DEFINITION. *Let α be a norm on $X \otimes Y$. We say that α is a crossnorm if, for all $x \in X$ and $y \in Y$,*

$$\alpha(x \otimes y) = \|x\| \, \|y\|.$$

We say that α is a reasonable norm if, for all $\phi \in X^*$ and $\psi \in Y^*$, the linear form $\phi \otimes \psi$ is bounded on $(X \otimes Y, \alpha)$ and has norm equal to $\|\phi\| \, \|\psi\|$.

1.3 DEFINITION. *Let α be a norm on $X \otimes Y$. Define α^* on $X^* \otimes Y^*$ by the equation*

$$\alpha^*\left(\sum_{i=1}^{n} \phi_i \otimes \psi_i \right) = \sup \left\{ \sum_{i=1}^{n} \sum_{j=1}^{m} \phi_i(x_j)\psi_i(y_j) : \alpha\left(\sum_{j=1}^{m} x_j \otimes y_j \right) = 1 \right\}.$$

We admit the possibility that α^* may take $+\infty$ as a value.

Observe that if α is reasonable then α^* is the norm induced on $X^* \otimes Y^*$ by considering the latter as a linear subspace of $(X \otimes Y, \alpha)^*$. The norm α^* is the **associate** of α.

1.4 LEMMA. *Let α be a norm on $X \otimes Y$ satisfying*

(i) $\alpha(x \otimes y) \leq \|x\| \, \|y\|$ for all $x \in X$ and $y \in Y$;

(ii) $\alpha^*(\phi \otimes \psi) \leq \|\phi\| \, \|\psi\|$ for all $\phi \in X^*$ and $\psi \in Y^*$.

Then α is a reasonable crossnorm.

PROOF. If $x \in X$, $y \in Y$, $\phi \in X^*$, and $\psi \in Y^*$, then

$$\phi(x)\psi(y) = (\phi \otimes \psi)(x \otimes y) \leq \alpha^*(\phi \otimes \psi)\alpha(x \otimes y) \leq \begin{cases} \|\phi\| \, \|\psi\| \alpha(x \otimes y) \\ \|x\| \, \|y\| \alpha^*(\phi \otimes \psi). \end{cases}$$

In this inequality, take a supremum as ϕ and ψ range over the unit cells of X^* and Y^*, getting

$$\|x\| \, \|y\| \leq \alpha(x \otimes y).$$

4

If we take a supremum as x and y range over the unit cells of X and Y, we have

$$\|\phi\|\,\|\psi\| \leq \alpha^*(\phi \otimes \psi). \quad \blacksquare$$

1.5 LEMMA. *Suppose that α is a reasonable crossnorm on $X \otimes Y$. Then α^* is a reasonable crossnorm on $X^* \otimes Y^*$.*

PROOF. Since α is a reasonable norm, α^* is a crossnorm. In order to show that α^* is a reasonable norm, take $\mu \in X^{**}$ and $\nu \in Y^{**}$. By 1.4 we need only show that

$$\alpha^{**}(\mu \otimes \nu) \leq \|\mu\|\,\|\nu\|.$$

By Goldstine's theorem [**57**, p. 424] there exist nets

$$\{x_\beta\} \subset X, \quad \{y_\gamma\} \subset Y$$

such that μ and ν are the weak* limits of $\{x_\beta\}$ and $\{y_\gamma\}$, and

$$\|x_\beta\| \leq \|\mu\|, \quad \|y_\gamma\| \leq \|\nu\|.$$

Let

$$\theta = \sum_{i=1}^{n} \phi_i \otimes \psi_i \in X^* \otimes Y^*.$$

Then

$$|\theta(x_\beta \otimes y_\gamma)| \leq \alpha^*(\theta)\alpha(x_\beta \otimes y_\gamma) \leq \alpha^*(\theta)\|x_\beta\|\,\|y_\gamma\| \leq \alpha^*(\theta)\|\mu\|\,\|\nu\|.$$

By taking a limit in β and γ we get

$$|\theta(\mu \otimes \nu)| \leq \alpha^*(\theta)\|\mu\|\,\|\nu\|$$
$$\text{or} \quad |(\mu \otimes \nu)(\theta)| \leq \alpha^*(\theta)\|\mu\|\,\|\nu\|.$$

This shows that $\alpha^{**}(\mu \otimes \nu) \leq \|\mu\|\,\|\nu\|$. $\quad \blacksquare$

Returning now to our only example thus far of a crossnorm, we refer to λ as the least of the reasonable crossnorms by virtue of the following lemma.

1.6 LEMMA. *The norm λ is a reasonable crossnorm on $X \otimes Y$. Moreover, if α is any reasonable crossnorm on $X \otimes Y$ then*

$$\lambda(z) \leq \alpha(z) \quad \text{for all} \quad z \in X \otimes Y.$$

PROOF. In our preliminary discussion of λ we verified the crossnorm property. We have

$$(1) \qquad \lambda\left(\sum_{i=1}^{n} x_i \otimes y_i\right) = \sup\left\{ \left\|\sum_{i=1}^{n} \phi(x_i)y_i\right\| : \phi \in X^*, \|\phi\| = 1 \right\}.$$

5

If now $\phi \in X^*$, $\psi \in Y^*$ and z is any element of $X \otimes Y$ then by Eq. (1)

$$|(\phi \otimes \psi)(z)| = \|\phi\| \|\psi\| \left| \left(\frac{\phi}{\|\phi\|} \otimes \frac{\psi}{\|\psi\|} \right)(z) \right|$$
$$\leq \|\phi\| \|\psi\| \lambda(z).$$

Thus $\lambda^*(\phi \otimes \psi) \leq \|\phi\| \|\psi\|$ and so by 1.4 λ is a reasonable crossnorm.

Now let α be a reasonable crossnorm on $X \otimes Y$. Let $z \in X \otimes Y$, $\phi \in X^*$ and $\psi \in Y^*$. Then

$$|(\phi \otimes \psi)(z)| \leq \alpha^*(\phi \otimes \psi)\alpha(z) = \|\phi\| \|\psi\| \alpha(z).$$

Taking a supremum for ϕ, ψ belonging to the unit cells of X^* and Y^* and using Eq. (1) gives $\lambda(z) \leq \alpha(z)$. ∎

It is also possible to define the greatest crossnorm. We do this next and then proceed to investigate each of these norms in somewhat greater detail.

1.7 DEFINITION. *For any $z \in X \otimes Y$ we define*

$$\gamma(z) = \inf \left\{ \sum_{i=1}^{n} \|x_i\| \|y_i\| : x_i \in X, \ y_i \in Y, \ z = \sum_{i=1}^{n} x_i \otimes y_i \right\}.$$

1.8 LEMMA. *The norm γ is a reasonable crossnorm on $X \otimes Y$. Furthermore, if α is any crossnorm on $X \otimes Y$ then $\gamma(z) \geq \alpha(z)$ for all $z \in X \otimes Y$.*

PROOF. The first issue at stake here is whether γ is a norm at all. We check first that if $z \neq 0$ then $\gamma(z) > 0$. Let

$$z = \sum_{i=1}^{n} x_i \otimes y_i.$$

If z is nonzero then the operator L_z defined by z is not the zero operator from X^* into Y. Hence there exists a $\phi \in X^*$ such that $\|\phi\| = 1$ and $L_z\phi \neq 0$. Thus

$$0 < \|L_z\phi\| \leq \left\| \sum_{i=1}^{n} \phi(x_i)y_i \right\| \leq \sum_{i=1}^{n} |\phi(x_i)| \|y_i\| \leq \sum_{i=1}^{n} \|x_i\| \|y_i\|.$$

Since this holds for *all* representations of z we have $\gamma(z) > 0$. The definition of $\gamma(z)$ is clearly independent of the representation of $z \in X \otimes Y$ and the remaining properties of a norm are easily verified.

To establish that γ is a crossnorm, suppose that $x \otimes y$ and $\sum_{i=1}^{n} x_i \otimes y_i$ are two equivalent expressions for $z \in X \otimes Y$. Then for all $\phi \in X^*$ we have

$$\phi(x)y = \sum_{i=1}^{n} \phi(x_i)y_i.$$

6

Now choose $\phi \in X^*$ so that $\|\phi\| = 1$ and $\phi(x) = \|x\|$. Then

$$\|x\|\,\|y\| = \|\phi(x)y\| = \left\|\sum_{i=1}^{n}\phi(x_i)y_i\right\| \leq \sum_{i=1}^{n}\|x_i\|\,\|y_i\|.$$

Hence $\gamma(z) = \|x\|\,\|y\|$.

To show that γ is a reasonable norm, take

$$z = \sum_{i=1}^{n} x_i \otimes y_i \in X \otimes Y,$$

and $\phi \in X^*$, $\psi \in Y^*$. Then

$$|(\phi \otimes \psi)(z)| \leq \sum_{i=1}^{n}|\phi(x_i)|\,|\psi(y_i)|$$

$$\leq \|\phi\|\,\|\psi\|\sum_{i=1}^{n}\|x_i\|\,\|y_i\|.$$

Since this bound holds for all representations of z, we have

$$|(\phi \otimes \psi)(z)| \leq \|\phi\|\,\|\psi\|\gamma(z),$$

so that $\gamma^*(\phi \otimes \psi) \leq \|\phi\|\,\|\psi\|$. An application of 1.4 completes the argument.

Finally, we show that γ is the greatest crossnorm. Let α denote any crossnorm on $X \otimes Y$ and let

$$z = \sum_{i=1}^{n} x_i \otimes y_i \in X \otimes Y.$$

Then

$$\alpha(z) = \alpha\left(\sum_{i=1}^{n} x_i \otimes y_i\right) \leq \sum_{i=1}^{n}\alpha(x_i \otimes y_i) = \sum_{i=1}^{n}\|x_i\|\,\|y_i\|.$$

Taking an infimum over all representations of z, we get $\alpha(z) \leq \gamma(z)$. ∎

1.9 LEMMA. *Let X and Y be Banach spaces and suppose that α and β are two norms on $X \otimes Y$ with $\alpha \geq \beta$. Then $\alpha^* \leq \beta^*$.*

PROOF. Let $\theta \in X^* \otimes Y^*$ be represented by

$$\theta = \sum_{i=1}^{n}\phi_i \otimes \psi_i.$$

Then

$$\beta^*(\theta) = \sup\left\{\sum_{i=1}^{n}(\phi_i \otimes \psi_i)(z) : z \in X \otimes Y, \ \beta(z) \leq 1\right\}$$

$$\geq \sup\left\{\sum_{i=1}^{n}(\phi_i \otimes \psi_i)(z) : z \in X \otimes Y, \ \alpha(z) \leq 1\right\}$$

$$= \alpha^*(\theta). \ \blacksquare$$

7

1.10 LEMMA. *Let X and Y be Banach spaces and let α be a crossnorm on $X \otimes Y$. Then α^* is a crossnorm on $X^* \otimes Y^*$ if and only if $\alpha \geq \lambda$.*

PROOF. Suppose that $\alpha \geq \lambda$. Then 1.9 gives $\alpha^* \leq \lambda^*$ and so

$$\alpha^*(\phi \otimes \psi) \leq \lambda^*(\phi \otimes \psi) = \|\phi\| \, \|\psi\| \quad \text{for all} \quad \phi \in X^*, \ \psi \in Y^*.$$

Also

$$\|\phi\| \, \|\psi\| = \sup\{|\phi(x)| \, |\psi(y)| : x \in X, \ y \in Y, \ \|x\| = \|y\| = 1\}$$
$$\leq \sup\{\alpha^*(\phi \otimes \psi)\alpha(x \otimes y) : x \in X, \ y \in Y, \ \|x\| = \|y\| = 1\}$$
$$= \alpha^*(\phi \otimes \psi).$$

Conversely, if α^* is a crossnorm on $X^* \otimes Y^*$ then α is a reasonable crossnorm. Hence by 1.6 we have $\alpha \geq \lambda$. ∎

1.11 DEFINITION. *Let X and Y be Banach spaces. A crossnorm α on $X \otimes Y$ is said to be uniform if for every pair of operators, $A \in \mathcal{L}(X, X)$ and $B \in \mathcal{L}(Y, Y)$, we have*

$$\alpha\left(\sum_{i=1}^{n} Ax_i \otimes By_i\right) \leq \|A\| \, \|B\| \alpha\left(\sum_{i=1}^{n} x_i \otimes y_i\right) \quad \text{for all } x_i \in X, \ y_i \in Y.$$

Later in this chapter (1.29) the tensor product $A \otimes B$ of two operators will be defined. Then it will become clear that the uniformity of the crossnorm α is equivalent to the assertion that for all A and B, the norm of $A \otimes B$, as an operator on $(X \otimes Y, \alpha)$, is $\|A\| \, \|B\|$.

1.12 LEMMA. *Both λ and γ are uniform crossnorms on $X \otimes Y$.*

PROOF. Let $A \in \mathcal{L}(X, X)$, $B \in \mathcal{L}(Y, Y)$, $\phi \in X^*$, $\psi \in Y^*$ and $\|\phi\| = \|\psi\| = 1$. Then

$$\psi\left(\sum_{i=1}^{n} \phi(Ax_i)By_i\right) = \sum_{i=1}^{n} \phi(Ax_i)\psi(By_i)$$
$$= (A^*\phi \otimes B^*\psi)\left(\sum_{i=1}^{n} x_i \otimes y_i\right)$$
$$\leq \lambda^*(A^*\phi \otimes B^*\psi)\lambda\left(\sum_{i=1}^{n} x_i \otimes y_i\right)$$
$$= \|A^*\phi\| \, \|B^*\psi\|\lambda\left(\sum_{i=1}^{n} x_i \otimes y_i\right)$$
$$\leq \|A\| \, \|B\|\lambda\left(\sum_{i=1}^{n} x_i \otimes y_i\right).$$

If we take a supremum on ψ we get

$$\left\|\sum_{i=1}^{n} \phi(Ax_i)By_i\right\| \leq \|A\| \, \|B\|\lambda\left(\sum_{i=1}^{n} x_i \otimes y_i\right).$$

If we take a supremum on ϕ we get

$$\lambda\left(\sum_{i=1}^{n} Ax_i \otimes By_i\right) \leq \|A\|\,\|B\|\lambda\left(\sum_{i=1}^{n} x_i \otimes y_i\right).$$

For γ we take

$$z = \sum_{i=1}^{n} x_i \otimes y_i \in X \otimes Y.$$

Then

$$\gamma\left(\sum_{i=1}^{n} Ax_i \otimes By_i\right) \leq \sum_{i=1}^{n} \|Ax_i\|\,\|By_i\| \leq \|A\|\,\|B\| \sum_{i=1}^{n} \|x_i\|\,\|y_i\|.$$

By taking an infimum over all representations of z, we get

$$\gamma\left(\sum_{i=1}^{n} Ax_i \otimes By_i\right) \leq \|A\|\,\|B\|\gamma(z). \quad \blacksquare$$

In general, the space $X \otimes Y$ equipped with a reasonable crossnorm is not complete. We denote the completion of $X \otimes Y$ with respect to α by $X \otimes_\alpha Y$. One of the attractive features of tensor product theory is that familiar spaces can be realised as tensor products of more elementary spaces with respect to a suitable crossnorm. We provide some of these realisations next. When X and Y are Banach spaces which are isometrically isomorphic to each other, we write simply $X = Y$.

If S is a compact Hausdorff space and Y is a Banach space, $C(S,Y)$ denotes the Banach space of all continuous maps f from S into Y with norm defined by

$$\|f\| = \sup_{s \in S} \|f(s)\|.$$

1.13 THEOREM. *For any compact Hausdorff space S and any Banach space Y,*

$$C(S) \otimes_\lambda Y = C(S,Y).$$

PROOF. With each element

$$z = \sum_{i=1}^{n} x_i \otimes y_i$$

in $C(S) \otimes Y$ we associate an element F_z in $C(S,Y)$ by defining

$$F_z(s) = \sum_{i=1}^{n} x_i(s)y_i.$$

9

Observe that

$$\lambda(z) = \sup_{\substack{\psi \in Y^* \\ \|\psi\|=1}} \left\| \sum_{i=1}^{n} \psi_i(y_i)x_i \right\|$$

$$= \sup_{\psi} \sup_{s} \left| \sum_{i=1}^{n} \psi(y_i)x_i(s) \right|$$

$$= \sup_{s} \sup_{\psi} \left| \psi\left(\sum_{i=1}^{n} x_i(s)y_i \right) \right|$$

$$= \sup_{s} \| F_z(s) \|$$

$$= \| F_z \|.$$

Thus the linear map $z \mapsto F_z$, after being extended by continuity, is an isometry of $C(S) \otimes_\lambda Y$ into $C(S,Y)$. It remains to be proved that the image of $C(S) \otimes Y$ is dense in $C(S,Y)$. Let $f \in C(S,Y)$ and let $\epsilon > 0$. Since f is continuous and S is compact the set $K = f(S)$ is compact and hence totally bounded. Therefore there exist points $y_1, \ldots, y_n$ in K such that the open cells $B(y_i, \epsilon)$ cover K. By the theorem on partitions of unity [148, p. 41], there exist functions $g_1, \ldots, g_n$ in $C(K)$ such that

$$0 \le g_i \le 1, \quad g_i(v) = 0 \text{ when } v \in K \backslash B(y_i, \epsilon), \text{ and } \sum_{i=1}^{n} g_i = 1.$$

Define $x_i = g_i \circ f$. Then

$$x_i \in C(S), \quad 0 \le x_i \le 1 \text{ and } \sum_{i=1}^{n} x_i = 1.$$

Put $z = \sum_{i=1}^{n} x_i \otimes y_i$. If $s \in S$, then

$$\left\| f(s) - \sum_{i=1}^{n} x_i(s)y_i \right\| = \left\| \sum_{i=1}^{n} x_i(s)[f(s) - y_i] \right\|$$

$$\le \sum_{i=1}^{n} x_i(s)\| f(s) - y_i \| < \epsilon,$$

because $x_i(s) = 0$ when $\| f(s) - y_i \| \ge \epsilon$. This proves that $\| f - F_z \| < \epsilon$. ∎

1.14 COROLLARY. *Let S and T be compact Hausdorff spaces. Then*

$$C(S) \otimes_\lambda C(T) = C(S \times T).$$

PROOF. In the previous theorem we take $Y = C(T)$ and so obtain

$$C(S) \otimes_\lambda C(T) = C(S, C(T)).$$

10

It is now elementary to make the identification $C(S, C(T)) = C(S \times T)$. With $f \in C(S \times T)$ we associate

$$\tilde{f} \in C(S, C(T)) \quad \text{where} \quad \tilde{f}(s) = f_s$$

and f_s is a section of f defined by $f_s(t) = f(s,t)$. ∎

The next theorem refers to spaces of the type $L_1(S, Y)$, consisting of Bochner integrable functions defined on a measure space S and taking values in a Banach space Y. In Chapter 10, the theory of such spaces is outlined.

1.15 THEOREM. *For any measure space S and any Banach space Y,*

$$L_1(S) \otimes_\gamma Y = L_1(S, Y).$$

PROOF. With each element

$$z = \sum_{i=1}^{n} x_i \otimes y_i \quad \text{in} \quad L_1(S) \otimes Y$$

we associate a function $F_z : S \to Y$ by defining

$$F_z(s) = \sum_{i=1}^{n} x_i(s) y_i.$$

Since each x_i can be approximated in $L_1(S)$ to any given precision by simple functions, the same is true of F_z. As for the norms we have

$$\|F_z\| = \int \|F_z(s)\| ds = \int \left\| \sum_{i=1}^{n} x_i(s) y_i \right\| ds \leq \int \sum_{i=1}^{n} |x_i(s)| \, \|y_i\| ds$$

$$= \sum_{i=1}^{n} \int |x_i(s)| \, \|y_i\| ds = \sum_{i=1}^{n} \|x_i\| \, \|y_i\|.$$

This shows that $F_z \in L_1(S, Y)$. (Refer to Chapter 10.) By taking an infimum over all representations of the element z, we obtain $\|F_z\| \leq \gamma(z)$. The linear mapping $z \mapsto F_z$ thus has a continuous extension to $L_1(S) \otimes_\gamma Y$. In order to see that this map is an isometry, we show that when F_z is a simple function, $\|F_z\| = \gamma(z)$. In this case, we can assume that

$$x_i x_j = 0 \quad \text{for} \quad i \neq j \quad \text{and} \quad \sum_{i=1}^{n} x_i = 1.$$

Then

$$\int \left\| \sum_{i=1}^{n} x_i(s) y_i \right\| ds = \sum_{i=1}^{n} \int |x_i(s)| \, \|y_i\| ds = \sum_{i=1}^{n} \|x_i\| \, \|y_i\| \geq \gamma(z).$$

Since the simple functions of the form F_z are dense in $L_1(S, Y)$, this completes the proof. ∎

Using the Fubini theorem and the ideas of 1.14 we have the following result:

1.16 COROLLARY. *If S and T are σ-finite measure spaces, then*

$$L_1(S) \otimes_\gamma L_1(T) = L_1(S \times T).$$

Let α be a norm on $X \otimes Y$, and let G and H be subspaces of X and Y respectively. Then $G \otimes H$ is a subspace of $X \otimes Y$ which inherits the natural norm $\alpha \mid G \otimes H$. The completion of $G \otimes H$ with this norm is $G \otimes_{\alpha|G\otimes H} H$. This is nothing but the closure of $G \otimes H$ in $X \otimes_\alpha Y$. To avoid cumbersome notation, we write this closure as $G \bar\otimes H$. (The exact meaning of this notation will therefore depend on the context.)

For a norm α that can be defined on the tensor product of any pair of Banach spaces, it is possible for $G \bar\otimes H$ to differ from $G \otimes_\alpha H$. For example, if $\alpha = \gamma$ and $z \in G \otimes H$, then we have

$$(\gamma \mid G \otimes H)(z) = \inf\left\{ \sum_{i=1}^n \|x_i\|\,\|y_i\| : z = \sum_{i=1}^n x_i \otimes y_i,\ x_i \in X, y_i \in Y \right\}$$

while

$$\gamma(z) = \inf\left\{ \sum_{i=1}^n \|g_i\|\,\|h_i\| : z = \sum_{i=1}^n g_i \otimes h_i,\ g_i \in G, h_i \in H \right\}.$$

Therefore $\gamma(z) \geq (\gamma \mid G \otimes H)(z)$ for $z \in G \otimes H$. In general, a strict inequality will hold here, and $G \bar\otimes H$ will be different from $G \otimes_\gamma H$. See [55, p. 227]. The next theorem addresses this question for the λ-norm.

1.17 THEOREM. *Let U be a closed subspace of the Banach space X and let Y be a Banach space. Then $U \otimes_\lambda Y$ is a closed subspace of $X \otimes_\lambda Y$.*

PROOF. Let us adopt temporarily the notation $\lambda_{U\otimes Y}$ and $\lambda_{X\otimes Y}$ for the λ-norms of an element z when thought of as a member of $U \otimes Y$ or $X \otimes Y$ respectively. It will suffice to show that if z is in the linear space $U \otimes Y$ then

$$\lambda_{U\otimes Y}(z) = \lambda_{X\otimes Y}(z).$$

If $z = \sum_{i=1}^n u_i \otimes y_i$, then

$$\lambda_{U\otimes Y}(z) = \sup\left\{ \left\| \sum_{i=1}^n \phi(u_i)y_i \right\| : \phi \in U^*,\ \|\phi\| = 1 \right\}.$$

By using the Hahn-Banach theorem to extend each $\phi \in U^*$ to a $\bar\phi \in X^*$ with $\|\bar\phi\| = \|\phi\|$ we see that

$$\lambda_{U\otimes Y}(z) \leq \lambda_{X\otimes Y}(z).$$

By restricting functionals in X^* to act only on U we obtain the reverse inequality. ∎

1.18 LEMMA. *Let X and Y be Banach spaces. If $z \in X \otimes_\gamma Y$ and $\epsilon > 0$ then there exist $x_n \in X$ and $y_n \in Y$ such that*

(i)
$$z = \sum_{n=1}^{\infty} x_n \otimes y_n,$$

(ii)
$$\|x_n\| = \|y_n\| \to 0,$$

(iii)
$$\gamma(z) \le \sum_{n=1}^{\infty} \|x_n\|\,\|y_n\| \le \gamma(z) + \epsilon.$$

PROOF. Select $z_n \in X \otimes Y$ such that

$$\gamma(z - z_n) < \epsilon/2^{n+2}, \quad \text{for} \quad n = 1, 2, \ldots.$$

From the definition of γ, we can write

$$z_1 = \sum_{i=1}^{k(1)} x_i \otimes y_i \quad \text{where} \quad \sum_{i=1}^{k(1)} \|x_i\|\,\|y_i\| < \gamma(z_1) + \epsilon/4 < \gamma(z) + \epsilon/2.$$

It is clear that here (and below) we can assume $\|x_i\| = \|y_i\|$. Now for each n,

$$\gamma(z_{n+1} - z_n) \le \gamma(z_{n+1} - z) + \gamma(z - z_n) \le \epsilon/2^{n+3} + \epsilon/2^{n+2} < \epsilon/2^{n+1}.$$

Thus we can represent $z_{n+1} - z_n$ as follows:

$$z_{n+1} - z_n = \sum_{i=k(n)+1}^{k(n+1)} x_i \otimes y_i \quad \text{with} \quad \sum_{i=k(n)+1}^{k(n+1)} \|x_i\|\,\|y_i\| < \epsilon/2^{n+1}.$$

The series $\sum_{i=1}^{\infty} x_i \otimes y_i$ is absolutely convergent in $X \otimes_\gamma Y$ since

$$\sum_{i=1}^{\infty} \gamma(x_i \otimes y_i) = \sum_{i=1}^{\infty} \|x_i\|\,\|y_i\| = \sum_{i=1}^{k(1)} \|x_i\|\,\|y_i\| + \sum_{n=1}^{\infty} \sum_{i=k(n)+1}^{k(n+1)} \|x_i\|\,\|y_i\|$$

$$< \gamma(z) + \epsilon/2 + \sum_{n=1}^{\infty} \epsilon/2^{n+1} = \gamma(z) + \epsilon.$$

The series $\sum_{i=1}^{\infty} x_i \otimes y_i$ is therefore convergent in $X \otimes_\gamma Y$. Its limit is z, since

$$\gamma\left[\sum_{i=1}^{k(n)} x_i \otimes y_i - z \right] = \gamma[z_1 + (z_2 - z_1) + \cdots + (z_n - z_{n-1}) - z]$$

$$= \gamma(z_n - z) \to 0.$$

It follows that

$$\gamma(z) \le \sum_{i=1}^{\infty} \gamma(x_i \otimes y_i) = \sum_{i=1}^{\infty} \|x_i\|\,\|y_i\| < \gamma(z) + \epsilon.$$

13

Since $\|x_i\|\,\|y_i\| \to 0$ and $\|x_i\| = \|y_i\|$, we must have $\|x_i\| \to 0$ and $\|y_i\| \to 0$. ■

1.19 DEFINITION. *Let X and Y be Banach spaces, and let α be a norm on $X \otimes Y$. Let $A \in \mathcal{L}(X, Y^*)$. If the supremum*

$$\sup\left\{ \left| \sum_{i=1}^{n} (Ax_i)(y_i) \right| : \alpha\left(\sum_{i=1}^{n} x_i \otimes y_i \right) = 1 \right\}$$

is finite, we denote it by $\|A\|_\alpha$. The set of all operators having $\|A\|_\alpha < \infty$ is denoted by $\mathcal{L}_\alpha(X, Y^*)$.

In this definition the question of the independence of the value of $\sum_{i=1}^{n} (Ax_i)(y_i)$ over all possible expressions representing an element $z \in X \otimes Y$ does not arise. However it becomes important in the next theorem and so we establish this fact now.

1.20 LEMMA. *Let X, Y and Z be Banach spaces. If*

$$\sum_{i=1}^{n} x_i \otimes y_i \sim 0 \otimes 0$$

in $X \otimes Y$ then

$$\sum_{i=1}^{n} (Ax_i)(y_i) = 0$$

for all $A \in \mathcal{L}(X, \mathcal{L}(Y, Z))$.

PROOF. By the definition of equivalence of expressions,

$$\sum_{i=1}^{n} \psi(y_i) x_i = 0$$

for all $\psi \in Y^*$. If $\{x_1, \ldots, x_n\}$ is linearly independent, then $\psi(y_i) = 0$ for $1 \le i \le n$ and for all $\psi \in Y^*$. Hence $y_i = 0$ for $i = 1, \ldots, n$ and

$$\sum_{i=1}^{n} (Ax_i)(y_i) = 0.$$

If $\{x_1, \ldots, x_n\}$ is linearly dependent then, say,

$$x_n = \sum_{i=1}^{n-1} \lambda_i x_i.$$

As in the proof of 1.1, we have

$$\sum_{i=1}^{n} x_i \otimes y_i \sim \sum_{i=1}^{n-1} x_i \otimes (y_i + \lambda_i y_n) \sim 0 \times 0.$$

14

Also, by similar algebraic manipulations

$$\sum_{i=1}^{n}(Ax_i)(y_i) = \sum_{i=1}^{n-1}(Ax_i)(y_i + \lambda_i y_n).$$

If $\{x_1, \ldots, x_{n-1}\}$ is linearly independent, then as above we conclude that

$$y_i + \lambda_i y_n = 0 \quad \text{for} \quad 1 \le i \le n-1 \quad \text{and that} \quad \sum_{i=1}^{n}(Ax_i)(y_i) = 0.$$

By continuing the above arguments we either establish that

$$\sum_{i=1}^{n}(Ax_i)(y_i) = 0$$

or arrive eventually at a dyad $u \otimes v$ such that

$$u \otimes v \sim \sum_{i=1}^{n} x_i \otimes y_i \sim 0 \times 0,$$

and

$$(Au)(v) = \sum_{i=1}^{n}(Ax_i)(y_i).$$

Since either $u = 0$ or $v = 0$, we have

$$\sum_{i=1}^{n}(Ax_i)(y_i) = 0. \quad \blacksquare$$

1.21 THEOREM. *If X and Y are Banach spaces, and if α is a crossnorm on $X \otimes Y$, then*

$$(X \otimes_\alpha Y)^* = \mathcal{L}_\alpha(X, Y^*).$$

PROOF. Corresponding to an element $\theta \in (X \otimes_\alpha Y)^*$, define an element

$$A_\theta \in \mathcal{L}(X, Y^*)$$

by setting

$$(A_\theta x)(y) = \theta(x \otimes y).$$

It is clear that $A_\theta x$ is a linear functional and that A_θ is a linear operator. That $A_\theta x \in Y^*$ follows from the inequality

$$|(A_\theta x)(y)| \le \|\theta\| \alpha(x \otimes y) = \|\theta\| \, \|x\| \, \|y\|.$$

15

In order to see that $A_\theta \in \mathcal{L}_\alpha(X, Y^*)$ we write

$$\left| \sum_{i=1}^n (A_\theta x_i)(y_i) \right| = \left| \sum_{i=1}^n \theta(x_i \otimes y_i) \right| = \left| \theta\left(\sum_{i=1}^n x_i \otimes y_i \right) \right|.$$

By taking a supremum over all expressions of unit α-norm we obtain $\|A_\theta\|_\alpha = \|\theta\|$. Thus the mapping $\theta \mapsto A_\theta$ is an isometry of $(X \otimes_\alpha Y)^*$ into $\mathcal{L}_\alpha(X, Y^*)$. In order to see that this mapping is surjective, let B be an element of $\mathcal{L}_\alpha(X, Y^*)$. Define θ by

$$\theta\left(\sum_{i=1}^n x_i \otimes y_i \right) = \sum_{i=1}^n (Bx_i)(y_i).$$

The functional θ is well-defined by 1.20. The linearity of θ is clear, and $\theta \in (X \otimes_\alpha Y)^*$ since

$$\left| \theta\left(\sum_{i=1}^n x_i \otimes y_i \right) \right| = \left| \sum_{i=1}^n (Bx_i)(y_i) \right| \le \alpha\left(\sum_{i=1}^n x_i \otimes y_i \right) \|B\|_\alpha.$$

Finally, we note that $A_\theta = B$. ∎

1.22 COROLLARY. *Let X and Y be Banach spaces. Then*

$$(X \otimes_\gamma Y)^* = \mathcal{L}(X, Y^*).$$

PROOF. Using the preceding theorem we see that

$$(X \otimes_\gamma Y^*) = \mathcal{L}_\gamma(X, Y^*).$$

Thus we need only show that $\mathcal{L}_\gamma(X, Y^*) = \mathcal{L}(X, Y^*)$. If

$$A \in \mathcal{L}(X, Y^*) \quad \text{and} \quad z = \sum_{i=1}^n x_i \otimes y_i,$$

then

$$\left| \sum_{i=1}^n (Ax_i)(y_i) \right| \le \sum_{i=1}^n |(Ax_i)(y_i)| \le \sum_{i=1}^n \|Ax_i\| \|y_i\| \le \|A\| \sum_{i=1}^n \|x_i\| \|y_i\|.$$

Taking an infimum over all possible representations of $z \in X \otimes Y$, and using 1.20 we obtain

$$\left| \sum_{i=1}^n (Ax_i)(y_i) \right| \le \|A\| \gamma(z).$$

Taking a supremum over z with $\gamma(z) = 1$ gives $\|A\|_\gamma \le \|A\|$. For the reverse inequality take $x \in X$ such that $\|x\| = 1$ and $\|Ax\| \ge \|A\| - \epsilon$. Choose $y \in Y$, $\|y\| = 1$, such that $|(Ax)(y)| \ge \|Ax\| - \epsilon$. Then

$$\|A\| \le \|Ax\| + \epsilon \le |(Ax)(y)| + 2\epsilon \le \|A\|_\gamma + 2\epsilon. \quad \blacksquare$$

Our next objective is to prove similar theorems involving λ (the least of the reasonable crossnorms). Some preliminary definitions and results are needed before proceeding to these theorems.

1.23 DEFINITION. *A Banach space* X *has the approximation property if for each compact set* $K \subset X$ *and for each* $\epsilon > 0$ *there is a continuous finite-rank operator* $A : X \to X$ *such that for all* $x \in K$, $\|Ax - x\| < \epsilon$.

1.24 THEOREM. *Let* X *be a Banach space such that* X^* *has the approximation property. Then for any Banach space* Y, *the set of compact operators in* $\mathcal{L}(X,Y)$ *is the closure of the set of operators having finite rank.*

PROOF. Suppose that A is compact; then its adjoint A^* is also compact; see [57, p. 485]. Take $\epsilon > 0$ and let

$$K = \{A^*\psi : \psi \in Y^*, \ \|\psi\| \le 1\}.$$

Then the closure of K is a compact set in X^*, and so there exists a finite rank operator $F \in \mathcal{L}(X^*, X^*)$ such that

(2) $$\|\phi - F\phi\| < \epsilon \quad \text{for all} \quad \phi \in K.$$

Now FA^* belongs to $\mathcal{L}(X^{**}, Y^{**})$ and has finite-dimensional range. Hence its adjoint $A^{**}F^*$ belongs to $\mathcal{L}(X^{**}, Y^{**})$ and has finite-dimensional range. From Eq. (2), if $x \in X$, $\|x\| = 1$, and $\phi \in K$, then

$$|\phi(x) - (F\phi)(x)| \le \epsilon.$$

If $\psi \in Y^*$ and $\|\psi\| \le 1$, then $A^*\psi \in K$ and

$$|(A^*\psi)(x) - (FA^*\psi)(x)| \le \epsilon.$$

If $J_X : X \to X^{**}$ denotes the natural embedding, then

$$|(A^{**}J_X x)(\psi) - (A^{**}F^*J_X x)(\psi)| \le \epsilon.$$

By taking a supremum on ψ we get

$$\|A^{**}J_X x - A^{**}F^*J_X x\| \le \epsilon.$$

Since A is compact,

$$A^{**}(X^{**}) \subset J_Y(Y)$$

by 11.12, and so

$$\|J_Y^{-1}A^{**}J_X x - J_Y^{-1}A^{**}F^*J_X x\| \le \epsilon,$$

or

$$\|Ax - J_Y^{-1}A^{**}F^*J_X x\| \leq \epsilon.$$

Taking a supremum on x yields

$$\|A - J_Y^{-1}A^{**}F^*J_X\| \leq \epsilon.$$

Thus $J_Y^{-1}A^{**}F^*J_X$ is the required operator of finite rank.

Conversely, if $\{F_n\}$ is a sequence of finite rank operators in $\mathcal{L}(X, Y)$ then each F_n is clearly compact. The set of compact operators is closed in the norm topology [57, p. 486], and so if $\|A - F_n\| \to 0$ as $n \to \infty$, A must also be compact. ∎

1.25 THEOREM. *Let X and Y be Banach spaces. Then $X^* \otimes_\lambda Y^*$ is (isometrically isomorphic to) the closure in the norm topology of the set of all operators from X to Y^* which are of finite rank.*

PROOF. Let $\phi_i \in X^*$ and $\psi_i \in Y^*$, for $1 \leq i \leq n$. The expression

$$\sum_{i=1}^n \phi_i \otimes \psi_i$$

defines an operator A from X to Y^* by the equation

$$Ax = \sum_{i=1}^n \phi_i(x)\psi_i \qquad (x \in X).$$

Equivalent expressions lead to identical operators A. It is clear that every operator of finite rank from X to Y^* can be obtained in this way from an expression in $X^* \otimes Y^*$. Now define the operator B from X^{**} to Y^* by the equation

$$Bp = \sum_{i=1}^n p(\phi_i)\psi_i \qquad (p \in X^{**}).$$

Furthermore,

$$A^*q = \sum_{i=1}^n q(\psi_i)\phi_i \qquad (q \in Y^{**})$$

and

$$A^{**}p = \sum_{i=1}^n p(\phi_i)J\psi_i \qquad (p \in X^{**})$$

where J is the canonical embedding of Y^* in Y^{***}. Hence $JB = A^{**}$ and $B = J^{-1}A^{**}$, whence

$$\|A\| = \|A^{**}\| = \|B\| = \lambda\left(\sum_{i=1}^n \phi_i \otimes \psi_i\right). \quad \blacksquare$$

18

1.26 THEOREM. *Let X be a Banach space such that X^* has the approximation property. Then for any Banach space Y, $X^* \otimes_\lambda Y^*$ is the space of all compact operators from X to Y^*.*

PROOF. Use 1.24 and 1.25. ∎

1.27 COROLLARY. *Let X and Y be Banach spaces, it being assumed that X^* has the approximation property. In order that*

$$X^* \otimes_\lambda Y^* = (X \otimes_\gamma Y)^*$$

it is necessary and sufficient that each element of $\mathcal{L}(X, Y^*)$ be compact.

PROOF. Use 1.26 and 1.22. ∎

1.28 COROLLARY. *Let p, q be real numbers satisfying $1 \le p < q < \infty$. Then*

$$\ell_{q'} \otimes_\lambda \ell_p = (\ell_q \otimes_\gamma \ell_{p'})^*, \quad \text{where} \quad \frac{1}{q'} + \frac{1}{q} = \frac{1}{p'} + \frac{1}{p} = 1.$$

PROOF. A classical theorem [**126**, p. 76] asserts that if $1 \le p < q < \infty$ then every member of $\mathcal{L}(\ell_q, \ell_p)$ is compact. Thus in the previous corollary we take $X = \ell_q$, $Y^* = \ell_p$ and use the familiar identification

$$\ell_p^* = \ell_{p'}, \quad 1 \le p < \infty. \quad ∎$$

1.29 DEFINITION. *Let $U, V, X,$ and Y be Banach spaces. If*

$$A \in \mathcal{L}(X, U) \quad \text{and} \quad B \in \mathcal{L}(Y, V)$$

then $A \otimes B$ is defined on $X \otimes Y$ by

$$(A \otimes B) \sum_{i=1}^{n} x_i \otimes y_i = \sum_{i=1}^{n} A x_i \otimes B y_i.$$

Observe that $(A \otimes B)(z)$ is independent of the representation of z. Indeed, if

$$z = \sum_{i=1}^{n} x_i \otimes y_i,$$

then by the definition of γ,

$$\gamma\left(\sum_{i=1}^{n} A x_i \otimes B y_i\right) \le \sum_{i=1}^{n} \|A x_i\| \, \|B y_i\| \le \|A\| \, \|B\| \sum_{i=1}^{n} \|x_i\| \, \|y_i\|.$$

By taking an infimum over the representations of z, we obtain

$$\gamma(A \otimes B)(z) \le \|A\|\,\|B\|\gamma(z).$$

Thus if

$$\sum_{i=1}^{n} x_i \otimes y_i \sim 0 \otimes 0,$$

then $z = 0$ and $(A \otimes B)z = 0$.

1.30 LEMMA. *Let X and Y be Banach spaces, and let α be a uniform crossnorm on $X \otimes Y$. If $A \in \mathcal{L}(X, X)$ and $B \in \mathcal{L}(Y, Y)$, then $A \otimes B$ has a unique continuous linear extension, $A \otimes_\alpha B$, on $X \otimes_\alpha Y$.*

PROOF. By the uniform property of α,

$$\alpha\left(\sum_{i=1}^{n} Ax_i \otimes By_i\right) \le \|A\|\,\|B\|\alpha\left(\sum_{i=1}^{n} x_i \otimes y_i\right).$$

This establishes that $\|A \otimes B\| \le \|A\|\,\|B\|$. (By considering appropriate dyads, one can see that equality holds here.) By the standard theorem on extending bounded operators defined on dense subsets [165] we obtain the extension, denoted by $A \otimes_\alpha B$, and defined on $X \otimes_\alpha Y$. Its norm is $\|A\|\,\|B\|$. ∎

Let X and Y be real Hilbert spaces, in which the inner products are denoted by $\langle a, b \rangle$. We shall now consider the problem of constructing an inner product on $X \otimes Y$. A pseudo inner product is defined by

$$\langle x \otimes y, \ a \otimes b \rangle = \langle x, a \rangle\,\langle y, b \rangle.$$

This is extended by linearity, so that if

$$u = \sum_{i=1}^{n} x_i \otimes y_i \quad \text{and} \quad v = \sum_{j=1}^{m} a_j \otimes b_j$$

then

(3) $$\langle u, v \rangle = \sum_{i,j} \langle x_i, a_j \rangle\,\langle y_i, b_j \rangle.$$

The next three lemmas establish that Eq. (3) defines a genuine inner product on $X \otimes Y$.

1.31 LEMMA. *If $u \sim u'$ and $v \sim v'$ then*

$$\langle u', v' \rangle = \langle u, v \rangle.$$

20

PROOF. We prove only $\langle u, v \rangle = \langle u', v \rangle$. A similar proof shows that $\langle u', v \rangle = \langle u', v' \rangle$. Suppose

$$u = \sum_{i=1}^{n} x_i \otimes y_i, \quad u' = \sum_{i=1}^{k} x_i' \otimes y_i' \text{ and } v = \sum_{i=1}^{m} a_i \otimes b_i.$$

Since $u \sim u'$,

$$\sum_{i=1}^{n} \phi(x_i) y_i = \sum_{i=1}^{k} \phi(x_i') y_i'$$

for all $\phi \in X^*$. Hence

$$\sum_{i=1}^{n} \langle a_j, x_i \rangle y_i = \sum_{i=1}^{k} \langle a_j, x_i' \rangle y_i' \quad \text{for} \quad 1 \le j \le m$$

so that

$$\sum_{j=1}^{m} \sum_{i=1}^{n} \langle a_j, x_i \rangle \langle b_j, y_i \rangle = \sum_{j=1}^{m} \sum_{i=1}^{k} \langle a_j, x_i' \rangle \langle b_j, y_i' \rangle$$

and $\langle u, v \rangle = \langle u', v \rangle$. ∎

1.32 LEMMA. *Equation (3) defines a positive semi-definite form on $X \otimes Y$.*

PROOF. We invoke the theorem [18, p. 422] that a positive linear operator on a Hilbert space has a unique positive square root. Consider the matrix A whose elements are

$$a_{ij} = \langle x_i, x_j \rangle \quad \text{where} \quad x_i \in X.$$

This matrix is positive semi-definite since

$$\sum_{i,j=1}^{n} \lambda_i \langle x_i, x_j \rangle \lambda_j = \left\langle \sum_{i=1}^{n} \lambda_i x_i, \sum_{j=1}^{n} \lambda_j x_j \right\rangle \ge 0$$

for all vectors $(\lambda_1, \ldots, \lambda_n)$. The matrix A can then be written in a unique way as $A = B^2$, where B is also positive semi-definite. Since A is symmetric, we have $A = A^T = (B^T)^2$ and conclude (from the uniqueness of B) that $B = B^T$. Now we write

$$u = \sum_{i=1}^{n} x_i \otimes y_i \quad \text{and} \quad A = BB^T$$

so that

$$\langle u, u \rangle = \sum_{ij} \langle x_i, x_j \rangle \langle y_i, y_j \rangle = \sum_{ij} a_{ij} \langle y_i, y_j \rangle$$

$$= \sum_{ij} \sum_{k} b_{ik} b_{jk} \langle y_i, y_j \rangle$$

$$= \sum_{k} \left\langle \sum_{i} b_{ik} y_i, \sum_{j} b_{jk} y_j \right\rangle \ge 0. \quad ∎$$

1.33 LEMMA. *For an expression* $u = \Sigma x_i \otimes y_i$, *the following properties are equivalent:*

(i) $\langle u, v \rangle = 0$ *for all* $v \in X \otimes Y$;

(ii) $\langle u, u \rangle = 0$;

(iii) $u = 0$ *as an operator from* X *to* Y.

PROOF. The implication (i) $\Longrightarrow$ (ii) is trivial. For the implication (ii) $\Longrightarrow$ (iii), use the Cauchy-Schwarz inequality to conclude that $\langle u, v \rangle = 0$ for all $v \in X \otimes Y$. If $v = x \otimes b$ then

$$0 = \langle u, v \rangle = \Sigma \langle x_i, x \rangle \, \langle y_i, b \rangle = \langle \Sigma \langle x_i, x \rangle y_i, b \rangle.$$

Since b is arbitrary, $\Sigma \langle x_i, x \rangle y_i = 0$, showing that u is the zero operator. For the implication (iii) $\Longrightarrow$ (i), suppose $u = 0$ as an operator. That is, $u \sim 0$. Taking $u' = 0$ in 1.31, we conclude that $\langle u, v \rangle = 0$ for all v. ∎

Equation (3) clearly defines a linear and symmetric bilinear form on $X \otimes Y$. By 1.32 and 1.33, this bilinear form is a genuine inner product. The norm induced by this inner product is

$$\beta(u) = \sqrt{\langle u, u \rangle};$$

i.e.,

$$\beta(\Sigma x_i \otimes y_i) = \left\{ \sum_{ij} \langle x_i, x_j \rangle \, \langle y_i, y_j \rangle \right\}^{1/2}.$$

The completion of $X \otimes Y$ with norm β is denoted by $X \otimes_\beta Y$ and is, of course, a Hilbert space.

1.34 LEMMA. *The norm* β *is a reasonable crossnorm, and* $\beta^* = \beta$.

PROOF. The crossnorm property follows at once from the definition, since

$$\beta(x \otimes y) = [\langle x, x \rangle \, \langle y, y \rangle]^{1/2} = \|x\| \, \|y\|.$$

The norm β^*, as defined in 1.3, is in this case

$$\beta^* \left(\sum_{i=1}^{n} x_i \otimes y_i \right) = \sup \frac{\sum_{i=1}^{n} \sum_{j=1}^{m} \langle x_i, a_j \rangle \, \langle y_i, b_j \rangle}{\beta \left(\sum_{j=1}^{m} a_j \otimes b_j \right)}.$$

If

$$u = \sum_{i=1}^{n} x_i \otimes y_i \quad \text{and} \quad v = \sum_{j=1}^{m} a_j \otimes b_j$$

then

$$\beta^*(u) = \sup \frac{\langle u, v \rangle}{\beta(v)} = \beta(u). \quad \blacksquare$$

1.35 LEMMA. *For any operator* $L : X \to X$ *we have*

$$\beta\left(\sum_{i=1}^{n} Lx_i \otimes y_i \right) \leq \|L\|\beta\left(\sum_{i=1}^{n} x_i \otimes y_i \right).$$

PROOF. Define the $n \times n$ matrix $A = (a_{ij})$ where $a_{ij} = \langle y_i, y_j \rangle$. As in the proof of 1.32 A is positive semi-definite and has a factorization $A = BB^T$. Then we can write

$$\beta^2\left(\sum_i Lx_i \otimes y_i \right) = \sum_{ij} \langle Lx_i, Lx_j \rangle \langle y_i, y_j \rangle = \sum_{ij} a_{ij} \langle Lx_i, Lx_j \rangle$$

$$= \sum_{ij} \sum_{\nu} b_{i\nu} b_{j\nu} \langle Lx_i, Lx_j \rangle = \sum_{\nu} \left\langle \sum_i b_{i\nu} Lx_i, \sum_j b_{j\nu} Lx_j \right\rangle$$

$$= \sum_{\nu} \left\| \sum_i b_{i\nu} Lx_i \right\|^2 = \sum_{\nu} \left\| L \sum_i b_{i\nu} x_i \right\|^2 \leq \|L\|^2 \sum_{\nu} \left\| \sum_i b_{i\nu} x_i \right\|^2$$

$$= \|L\|^2 \sum_{\nu} \left\langle \sum_i b_{i\nu} x_i, \sum_j b_{j\nu} x_j \right\rangle = \|L\|^2 \sum_{ij} \sum_{\nu} b_{i\nu} b_{j\nu} \langle x_i, x_j \rangle$$

$$= \|L\|^2 \sum_{ij} a_{ij} \langle x_i, x_j \rangle = \|L\|^2 \beta^2\left(\sum_i x_i \otimes y_i \right). \quad \blacksquare$$

1.36 LEMMA. *The norm* β *is a uniform crossnorm.*

PROOF. Using the preceding lemma and a symmetry argument, we have for any two operators $L_1 : X \to X$ and $L_2 : Y \to Y$,

$$\beta\left(\sum_{i=1}^{n} L_1 x_i \otimes L_2 y_i \right) \leq \|L_1\|\beta\left(\sum_{i=1}^{n} x_i \otimes L_2 y_i \right) \leq \|L_1\| \|L_2\| \beta\left(\sum_{i=1}^{n} x_i \otimes y_i \right). \quad \blacksquare$$

1.37 LEMMA. *Let* G *and* H *be closed subspaces in Hilbert spaces* X *and* Y *respectively. Then the Hilbert-space crossnorm* β *on* $G \otimes H$ *is the restriction to* $G \otimes H$ *of the Hilbert-space crossnorm on* $X \otimes Y$.

PROOF. Denote the Hilbert-space crossnorm on $X \otimes Y$ by β. Thus

$$\beta(\Sigma x_i \otimes y_i) = \left\{ \sum_{ij} \langle x_i, x_j \rangle \langle y_i, y_j \rangle \right\}^{1/2} \quad x_i \in X, \ y_i \in Y.$$

The Hilbert-space crossnorm on $G \otimes H$ is defined by

$$\beta_1(\Sigma g_i \otimes h_i) = \left\{ \sum_{ij} \langle g_i, g_j \rangle \langle h_i, h_j \rangle \right\}^{1/2} \quad g_i \in G, \ h_i \in H.$$

23

It is therefore clear that β_1 is the restriction of β to $G \otimes H$. ■

1.38 LEMMA. *Let S and T be σ-finite measure spaces. Let*

$$\sum_{i=1}^{n} x_i \otimes y_i \in L_2(S) \otimes L_2(T)$$

and define

$$f(s,t) = \sum_{i=1}^{n} x_i(s) y_i(t).$$

If

$$\Sigma x_i \otimes y_i \sim 0 \otimes 0$$

then $f(s,t) = 0$ almost everywhere.

PROOF. Let μ and ν be the measures on S and T respectively. Let σ be the product measure, $\mu \otimes \nu$. By the Fubini Theorem [148, p. 150] we have then (with $z = \Sigma x_i \otimes y_i$)

$$\|f\|^2 = \int f^2 \, d\sigma = \iint f^2(s,t) \, ds \, dt = \iint \sum_{ij} x_i(s) y_i(t) x_j(s) y_j(t) \, ds \, dt$$

(4)
$$= \sum_{ij} \int x_i(s) x_j(s) \, ds \int y_i(t) y_j(t) \, dt$$

$$= \sum_{ij} \langle x_i, x_j \rangle \langle y_i, y_j \rangle = \langle z, z \rangle = \beta^2(z).$$

Now if

$$\sum_{i=1}^{n} x_i \otimes y_i \sim 0 \times 0,$$

then by 1.33, $\langle z, z \rangle = 0$. Hence $\|f\| = 0$, and $f(s,t) = 0$ almost everywhere. ■

1.39 THEOREM. *If S and T are σ-finite measure spaces, then*

$$L_2(S) \otimes_\beta L_2(T) = L_2(S \times T).$$

PROOF. Let z be a member of $L_2(S) \otimes L_2(T)$, say

$$z = \sum_{i=1}^{n} x_i \otimes y_i.$$

We associate with z the function f defined by

$$f(s,t) = \sum_{i=1}^{n} x_i(s) y_i(t).$$

24

By the preceding lemma, f does not depend on the representation chosen for z. Equation (4) in the preceding proof shows that $f \in L_2(S \times T)$. The map $z \mapsto f$ is linear, and by Eq. (4) it is norm-preserving. Thus the completion of $L_2(S) \otimes L_2(T)$ in the β-norm is its closure in $L_2(S \times T)$. To complete the proof we show that $L_2(S) \otimes L_2(T)$ has orthogonal complement 0 in $L_2(S \times T)$. To this end, let $f \in L_2(S \times T)$ and suppose that

$$\langle f, x \otimes y \rangle = 0 \quad \text{for all} \quad x \in L_2(S) \text{ and all } y \in L_2(T).$$

By the Fubini Theorem, we write this condition as

$$\int y(t) \left[\int x(s) f^t(s) \, ds \right] dt = 0.$$

Since y is arbitrary in $L_2(T)$ we conclude that for almost all t,

$$\int x(s) f^t(s) \, ds = 0.$$

Since x is arbitrary in $L_2(S)$, we conclude that for almost all t, f^t is the 0-element in $L_2(S)$. Hence

$$\int [f(s,t)]^2 ds = \int \left\{ \int [f^t(s)]^2 ds \right\} dt = 0. \quad \blacksquare$$

1.40 LEMMA. *Let X and Y be Hilbert spaces, and let P be the orthogonal projection of X onto a subspace G. Then $P \otimes_\beta I_Y$ is the orthogonal projection of $X \otimes_\beta Y$ onto $G \otimes_\beta Y$.*

PROOF. It is clear that $P \otimes_\beta I$ maps $X \otimes_\beta Y$ into $G \otimes_\beta Y$ and leaves invariant each element of $G \otimes_\beta Y$. For the orthogonality property we only need verify that

$$x \otimes y - (P \otimes_\beta I)(x \otimes y) \perp g \otimes v \quad (x \in X, \; y \in Y, \; g \in G, \; v \in Y).$$

Computing the inner-product, we get

$$\langle (x - Px) \otimes y, \; g \otimes v \rangle = \langle x - Px, g \rangle \langle y, v \rangle = 0$$

because P is the orthogonal projection of X onto G. $\quad \blacksquare$

1.41 DEFINITION. *If A and B belong to $\mathcal{L}(X, X)$ then their Boolean sums are defined by*

$$A \oplus B = A + B - AB$$

$$B \oplus A = B + A - BA.$$

1.42 THEOREM. *Let X and Y be two Hilbert spaces. Let P and Q be orthogonal projections on X and Y respectively. Then*

$$(P \otimes_\beta I) \oplus (I \otimes_\beta Q)$$

25

is the orthogonal projection of $X \otimes_\beta Y$ onto

$$\mathcal{R}(P) \otimes_\beta Y + X \otimes_\beta \mathcal{R}(Q).$$

PROOF. By 11.2, it remains only to prove that the given projection has the orthogonality property. It is to be proved that for arbitrary $z \in X \otimes_\beta Y$ and arbitrary

$$w \in \mathcal{R}(P) \otimes_\beta Y + X \otimes_\beta \mathcal{R}(Q)$$

we have

$$z - [(P \otimes I) \oplus (I \otimes Q)]z \perp w.$$

It suffices to prove this for $w \in \mathcal{R}(P) \otimes_\beta Y$ and then use a symmetry argument. Because of linearity and continuity it suffices to verify the orthogonality relationship for dyads:

$$z = x \otimes y \quad \text{and} \quad w = g \otimes v, \quad \text{where} \quad x \in X, \ y \in Y, \ g \in \mathcal{R}(P) \text{ and } v \in Y.$$

Calculating the required inner product, we have

$$\langle x \otimes y - Px \otimes y - x \otimes Qy + Px \otimes Qy, \ g \otimes v \rangle$$
$$= \langle x, g \rangle \langle y, v \rangle - \langle Px, g \rangle \langle y, v \rangle - \langle x, g \rangle \langle Qy, v \rangle + \langle Px, g \rangle \langle Qy, v \rangle$$
$$= \langle x - Px, g \rangle \langle y, v \rangle - \langle x - Px, g \rangle \langle Qy, v \rangle = 0$$

because $x - Px \perp g$. ∎

In previous theorems (*viz.* 1.14, 1.16, and 1.39) it has been shown that

(*i*) $$C(S) \otimes_\lambda C(T) = C(S \times T)$$

(*ii*) $$L_1(S) \otimes_\gamma L_1(T) = L_1(S \times T)$$

(*iii*) $$L_2(S) \otimes_\beta L_2(T) = L_2(S \times T).$$

In the remainder of this chapter, the analogous result for $L_p(S \times T)$, $1 < p < \infty$, will be developed.

1.43 DEFINITION. *Let Y be a Banach space. For $y_1, \ldots, y_n \in Y$ and $1 \le q < \infty$, we define*

$$\mu_q(y_1, \ldots, y_n) = \sup \left\{ \left(\sum_{i=1}^{n} |\psi(y_i)|^q \right)^{1/q} : \psi \in Y^*, \ \|\psi\| = 1 \right\}.$$

The case $q = \infty$ is given by

$$\mu_\infty(y_1, \ldots, y_n) = \sup \left\{ \max_{1 \le i \le n} |\psi(y_i)| : \psi \in Y^*, \ \|\psi\| = 1 \right\}.$$

26

1.44 LEMMA. *Let $1 \leq p, q \leq \infty$ and $p^{-1} + q^{-1} = 1$. Then*

$$\mu_q(y_1, \ldots, y_n) = \sup \left\{ \left\| \sum_{i=1}^n \lambda_i \, y_i \right\| : \left(\sum_{i=1}^n |\lambda_i|^p \right)^{1/p} \leq 1 \right\}.$$

PROOF. We fix the n-tuple $(y_1, \ldots, y_n)$, with $y_i \in Y$. Let

$$\lambda = (\lambda_1, \ldots, \lambda_n) \in \mathbf{R}^n, \quad \text{with} \quad \|\lambda\|_p = \left(\Sigma \, |\lambda_i|^p \right)^{1/p} = 1.$$

For each $\psi \in Y^*$ with $\|\psi\| = 1$, we set $\bar{\psi} = (\psi(y_1), \ldots, \psi(y_n))$. Thus $\bar{\psi} \in \mathbf{R}^n$. By using the duality between ℓ_p^n and ℓ_q^n, we have (from 1.43)

$$\mu_q(y_1, \ldots, y_n) = \sup_\psi \|\bar{\psi}\|_q = \sup_\psi \sup_\lambda \langle \lambda, \bar{\psi} \rangle$$

$$= \sup_\psi \sup_\lambda \sum_{i=1}^n \lambda_i \, \psi(y_i) = \sup_\lambda \sup_\psi \psi \left(\sum_{i=1}^n \lambda_i \, y_i \right)$$

$$= \sup_\lambda \left\| \sum_{i=1}^n \lambda_i \, y_i \right\|. \quad \blacksquare$$

1.45 DEFINITION. *Let $z \in X \otimes Y$, where X and Y are Banach spaces. For $1 \leq p \leq \infty$, the p-nuclear norm of z is defined by the following equation, in which $p^{-1} + q^{-1} = 1$:*

$$\alpha_p(z) = \inf \left\{ \left(\sum_{i=1}^n \|x_i\|^p \right)^{1/p} \mu_q(y_1, \ldots, y_n) \; : \; z = \sum_{i=1}^n x_i \otimes y_i \right\}.$$

The infimum is taken with respect to all representations of z, and $(\Sigma \, |\lambda_i|^p)^{1/p}$ is understood to mean $\max |\lambda_i|$ when $p = \infty$.

1.46 LEMMA. *The p-nuclear norm is a reasonable crossnorm.*

PROOF. Let $z \in X \otimes Y$, and let one of its representations be $z = \sum_{i=1}^n x_i \otimes y_i$. Let A be the operator from Y^* to X corresponding to z. Then for all ψ of norm 1 in Y^*, we have, by the Hölder inequality,

$$\|A\psi\| = \left\| \sum_{i=1}^n \psi(y_i) x_i \right\| \leq \sum_{i=1}^n |\psi(y_i)| \, \|x_i\|$$

$$\leq \left(\sum_{i=1}^n |\psi(y_i)|^q \right)^{1/q} \left(\sum_{i=1}^n \|x_i\|^p \right)^{1/p} \leq \mu_q(y_1, \ldots, y_n) \left(\sum_{i=1}^n \|x_i\|^p \right)^{1/p}.$$

In this inequality, take an infimum over all representations of z, followed by a supremum in ψ. The result is

$$\|A\| \leq \alpha_p(z).$$

27

This shows that if $z \neq 0$, then $A \neq 0$ and $\alpha_p(z) > 0$.

The triangle inequality for α_p is proved as follows. We want to show that for all $\epsilon > 0$,

$$\alpha_p(z + w) \leq \alpha_p(z) + \alpha_p(w) + 2\epsilon.$$

If ϵ is given, we select a representation for z, say

$$z = \sum_{i=1}^{n} x_i \otimes y_i,$$

such that

$$\left[\sum_{i=1}^{n} \|x_i\|^p \right]^{1/p} \mu_q(y_1, \ldots, y_n) \leq \alpha_p(z) + \epsilon.$$

Since $1/p + 1/q = 1$, and since the factors on the left in the preceding inequality are homogeneous, we can assume further that

$$\left[\sum_{i=1}^{n} \|x_i\|^p \right]^{1/p} \leq \left[\alpha_p(z) + \epsilon \right]^{1/p}$$

and that

$$\mu_q(y_1, \ldots, y_n) \leq [\alpha_p(z) + \epsilon]^{1/q}.$$

Then for all functionals ψ of norm 1 in Y^* we have

$$\left[\sum_{i=1}^{n} |\psi(y_i)|^q \right]^{1/q} \leq [\alpha_p(z) + \epsilon]^{1/q}.$$

Obviously then,

$$\sum_{i=1}^{n} \|x_i\|^p \leq \alpha_p(z) + \epsilon$$

and

$$\sum_{i=1}^{n} |\psi(y_i)|^q \leq \alpha_p(z) + \epsilon.$$

Now take a representation for w, say

$$w = \sum_{i=n+1}^{m} x_i \otimes y_i$$

with the analogous properties. Thus

$$\sum_{i=n+1}^{m} \|x_i\|^p \leq \alpha_p(w) + \epsilon$$

and

28

$$\sum_{i=n+1}^{m} |\psi(y_i)|^q \le \alpha_p(w) + \epsilon.$$

Then

$$\sum_{i=1}^{m} \|x_i\|^p \le \alpha_p(z) + \alpha_p(w) + 2\epsilon$$

and

$$\sum_{i=1}^{m} |\psi(y_i)|^q \le \alpha_p(z) + \alpha_p(w) + 2\epsilon.$$

Consequently

$$\left[\sum_{i=1}^{m} \|x_i\|^p\right]^{1/p} \left[\sum_{i=1}^{m} |\psi(y_i)|\right]^{1/q} \le \alpha_p(z) + \alpha_p(w) + 2\epsilon.$$

Upon taking a supremum in ψ, we obtain the desired inequality.

For $z = x \otimes y$, we have, immediately from 1.45,

$$(5) \qquad \alpha_p(x \otimes y) \le \left(\|x\|^p\right)^{1/p} \mu_q(y) = \|x\| \, \|y\|.$$

Now let $\phi \in X^*, \psi \in Y^*$, and $z = \sum_{i=1}^{n} x_i \otimes y_i$. Then

$$(\phi \otimes \psi)(z) = \sum_{i=1}^{n} \phi(x_i)\psi(y_i) \le \|\phi\| \sum_{i=1}^{n} \|x_i\| \, |\psi(y_i)|$$

$$\le \|\phi\| \left(\sum_{i=1}^{n} \|x_i\|^p\right)^{1/p} \left(\sum_{i=1}^{n} |\psi(y_i)|^q\right)^{1/q}$$

$$\le \|\phi\| \, \|\psi\| \left(\sum_{i=1}^{n} \|x_i\|^p\right)^{1/p} \mu_q(y_1, \ldots, y_n).$$

By taking an infimum over all the representations of z, we obtain

$$(\phi \otimes \psi)(z) \le \|\phi\| \, \|\psi\| \alpha_p(z).$$

By taking a supremum over all z for which $\alpha_p(z) \le 1$ we obtain, by 1.3,

$$(6) \qquad \alpha_p^*(\phi \otimes \psi) \le \|\phi\| \, \|\psi\|.$$

From Eqs. (5) and (6) and 1.4, α_p is a reasonable crossnorm. ∎

1.47 LEMMA. *Let Y be a Banach space, S a finite measure space, and $1 \le p \le \infty$. The natural embedding of $L_p(S) \otimes_{\alpha_p} Y$ into $L_p(S, Y)$ is of norm 1.*

PROOF. Let

$$z = \sum_{i=1}^{n} x_i \otimes y_i \in L_p(S) \otimes Y.$$

29

The natural embedding associates with z the function f defined by $f(s) = \sum_{i=1}^{n} x_i(s) y_i$. It is to be shown that $\|f\| \leq \alpha_p(z)$. If $1 \leq p < \infty$, then

$$
\begin{aligned}
\|f\|_p &= \left\{ \int_S \|f(s)\|^p \, ds \right\}^{1/p} = \left\{ \int_S \left\| \sum_{i=1}^{n} x_i(s) y_i \right\|^p ds \right\}^{1/p} \\
&= \left\{ \int_S \sup_\psi \left| \sum_{i=1}^{n} x_i(s) \psi(y_i) \right|^p ds \right\}^{1/p} \qquad (\psi \in Y^*, \ \|\psi\| = 1) \\
&\leq \left\{ \int_S \sup_\psi \sum_{i=1}^{n} |x_i(s)|^p \left[\sum_{j=1}^{n} |\psi(y_j)|^q \right]^{p/q} ds \right\}^{1/p} \\
&= \left\{ \sum_{i=1}^{n} \|x_i\|^p \right\}^{1/p} \mu_q(y_1, \ldots, y_n).
\end{aligned}
$$

For $p = \infty$, we write

$$
\begin{aligned}
\|f\|_\infty &= \operatorname*{ess\,sup}_s \|f(s)\| = \operatorname*{ess\,sup}_s \left\| \sum_{i=1}^{n} x_i(s) y_i \right\| \\
&= \operatorname*{ess\,sup}_s \sup_\psi \psi\left(\sum_{i=1}^{n} x_i(s) y_i \right) = \operatorname*{ess\,sup}_s \sup_\psi \sum_{i=1}^{n} x_i(s) \psi(y_i) \\
&\leq \operatorname*{ess\,sup}_s \sup_\psi \left\{ \max_i |x_i(s)| \right\} \sum_{i=1}^{n} |\psi(y_i)| \\
&= \left\{ \operatorname*{ess\,sup}_s \max_i |x_i(s)| \right\} \left\{ \sup_\psi \sum_{i=1}^{n} |\psi(y_i)| \right\} \\
&= \max_i \|x_i\| \, \mu_1(y_1, \ldots, y_n).
\end{aligned}
$$

Thus for $1 \leq p \leq \infty$ we have

$$
\|f\|_p \leq \left(\sum_{i=1}^{n} \|x_i\|^p \right)^{1/p} \mu_q(y_1, \ldots, y_n).
$$

By taking an infimum over all representations of z, we have by 1.45,

$$
\|f\|_p \leq \alpha_p(z).
$$

The embedding is the (unique) continuous extension of the map $z \mapsto f$ defined above. In order to see that the embedding is of norm 1, let z be a dyad, $z = x \otimes y$, in the preceding computation. One sees at once that

$$
\|f\| = \|x\| \, \|y\| = \alpha_p(z). \quad \blacksquare
$$

30

1.48 LEMMA. *If* $x_1, \ldots, x_n \in L_p(S)$, $1 \le p < \infty$ *and* $x_i x_j = 0$ *when* $i \ne j$, *then*

$$\mu_q(x_1, \ldots, x_n) = \max_i \|x_i\|.$$

PROOF. Let

$$E_i = \{s \in S : x_i(s) \ne 0\}.$$

Then by 1.44,

$$\mu_q(x_1, \ldots, x_n) = \sup_\lambda \left\{ \left\| \sum_{j=1}^n \lambda_j x_j \right\| \quad : \quad \|\lambda\|_p = 1 \right\}$$

$$= \sup_\lambda \left\{ \int \left| \sum_{j=1}^n \lambda_j x_j(s) \right|^p ds \right\}^{1/p}$$

$$= \sup_\lambda \left\{ \sum_{i=1}^n \int_{E_i} \left| \sum_{j=1}^n \lambda_j x_j(s) \right|^p ds \right\}^{1/p}$$

$$= \sup_\lambda \left\{ \sum_{i=1}^n \int_{E_i} |\lambda_i x_i(s)|^p ds \right\}^{1/p}$$

$$= \sup_\lambda \left\{ \sum_{i=1}^n |\lambda_i|^p \|x_i\|^p \right\}^{1/p}$$

$$= \sup_{\Sigma|\theta_i|=1} \left\{ \sum_{i=1}^n \theta_i \|x_i\|^p \right\}^{1/p}$$

$$= \{\max_i \|x_i\|^p\}^{1/p} = \max_i \|x_i\|. \quad \blacksquare$$

1.49 LEMMA. *Let Y be a Banach space, S a finite measure space, and $1 \le p < \infty$. The natural embedding of $L_p(S, Y)$ into $Y \otimes_{\alpha_p} L_p(S)$ is of norm 1.*

PROOF. Let f be a simple function in $L_p(S, Y)$. Then f can be written in the form

$$f(s) = \sum_{i=1}^n x_i(s) y_i,$$

where the functions x_i are multiples of characteristic functions. We can assume that $x_i x_j = 0$ if $i \ne j$ and that $\|x_i\| = 1$ for $1 \le i \le n$. The image of f under the natural embedding is the element $w = \sum_{i=1}^n y_i \otimes x_i$. It is to be shown that $\alpha_p(w) \le \|f\|$. By 1.48,

$$\mu_q(x_1, \ldots, x_n) = 1.$$

Hence by the definition of $\alpha_p(w)$ in 1.45,

$$\alpha_p(w) \le \left(\sum_{i=1}^n \|y_i\|^p \right)^{1/p}.$$

31

On the other hand, if $E_i = \{s : x_i(s) \neq 0\}$, then

$$\|f\| = \left\{ \int \|f(s)\|^p \, ds \right\}^{1/p}$$

$$= \left\{ \sum_{i=1}^n \int_{E_i} \left\| \sum_{j=1}^n x_j(s) y_j \right\|^p ds \right\}^{1/p}$$

$$= \left\{ \sum_{i=1}^n \int_{E_i} \| x_i(s) y_i \|^p \, ds \right\}^{1/p}$$

$$= \left\{ \sum_{i=1}^n \| y_i \|^p \int |x_i(s)|^p \, ds \right\}^{1/p}$$

$$= \left\{ \sum_{i=1}^n \| y_i \|^p \right\}^{1/p}.$$

Thus we conclude that $\alpha_p(w) \leq \|f\|$. The simple functions form a dense set in $L_p(S, Y)$ by [**57**, p. 125]. Hence the embedding (as defined thus far) has a unique linear extension of norm at most 1. If f in the previous argument is a constant function, say $f(s) = y$, then $w = y \otimes 1$ and

$$\alpha_p(w) = \|y\| \, \|1\| = \|f\|.$$

Hence the embedding is of norm 1. ∎

1.50 THEOREM. *Let S be a finite measure space, and $1 \leq p < \infty$. Let Y be a Banach space such that, under the natural map,*

$$L_p(S) \otimes_{\alpha_p} Y = Y \otimes_{\alpha_p} L_p(S).$$

Then $L_p(S) \otimes_{\alpha_p} Y = L_p(S, Y)$.

PROOF. Let Γ be the natural embedding of $L_p(S, Y)$ into $Y \otimes_{\alpha_p} L_p(S)$. By 1.49, $\|\Gamma\| = 1$. Let Φ be the natural map of $Y \otimes_{\alpha_p} L_p(S)$ onto $L_p(S) \otimes_{\alpha_p} Y$. By hypothesis, Φ is an isometry. Let Λ be the natural embedding of $L_p(S) \otimes_{\alpha_p} Y$ into $L_p(S, Y)$. By 1.47, $\|\Lambda\| = 1$. We will prove that

$$\Lambda \Phi \Gamma = I.$$

It suffices to prove that $\Lambda \Phi \Gamma f = f$ for any simple function f in $L_p(S, Y)$. Let $f(s) = \sum_{i=1}^n c_i(s) y_i$, where $y_i \in Y$ and the c_i are characteristic functions of measurable sets in S. Then

$$[\Lambda \Phi \Gamma f](s) = \left[\Lambda \Phi \left(\sum_{i=1}^n y_i \otimes c_i \right) \right](s) = \left[\Lambda \left(\sum_{i=1}^n c_i \otimes y_i \right) \right](s)$$

$$= \sum_{i=1}^n c_i(s) y_i = f(s).$$

The map $\Phi\Gamma$ is the required isometry, since

$$\|f\| = \|\Lambda\Phi\Gamma f\| \leq \|\Phi\Gamma f\| \leq \|f\|. \quad \blacksquare$$

Remark. Under the conditions of the theorem, the following diagram is commutative:

$$
\begin{array}{ccc}
L_p(S,Y) & \xrightarrow{\;\Gamma\;} & Y \otimes_{\alpha_p} L_p(S) \\[2mm]
\Big\downarrow{\scriptstyle I} & & \Big\downarrow{\scriptstyle \Phi} \\[2mm]
L_p(S,Y) & \xleftarrow{\;\Lambda\;} & L_p(S) \otimes_{\alpha_p} Y
\end{array}
$$

1.51 LEMMA. *Under the natural map, we have*

$$L_p(S) \otimes_{\alpha_p} L_p(T) = L_p(T) \otimes_{\alpha_p} L_p(S) \quad (1 \leq p < \infty).$$

PROOF. Let $z = \sum x_i \otimes y_i \in L_p(S) \otimes L_p(T)$. The natural map associates with z the element $u = \sum_{i=1}^{n} y_i \otimes x_i$ in $L_p(T) \otimes L_p(S)$. We want to show that

$$\alpha_p(z) = \alpha_p(u).$$

Because of symmetry, it suffices to prove that

$$\alpha_p(u) \leq \alpha_p(z).$$

With z, we associate $f \in L_p(S, L_p(T))$ by means of the natural map of 1.47. Thus, by 1.47,

$$f(s) = \sum_{i=1}^{n} x_i(s) y_i \quad \text{and} \quad \|f\| \leq \alpha_p(z).$$

With f, we associate $w \in L_p(T) \otimes_{\alpha_p} L_p(S)$ by means of the natural map of 1.49. Thus, by 1.49

$$w = \sum_{i=1}^{n} y_i \otimes x_i \quad \text{and} \quad \alpha_p(w) \leq \|f\|.$$

Since $w = u$, we have established that $\alpha_p(u) \leq \alpha_p(z)$. $\quad \blacksquare$

1.52 COROLLARY. For $1 \leq p < \infty$,

$$L_p(S) \otimes_{\alpha_p} L_p(T) = L_p(S \times T).$$

PROOF. By 1.51, the space $L_p(T)$ has the property required of Y in 1.50. Hence,

$$L_p(S) \otimes_{\alpha_p} L_p(T) = L_p(S, L_p(T)).$$

It only remains to prove that

$$L_p(S, L_p(T)) = L_p(S \times T).$$

With any $f \in L_p(S, L_p(T))$ we associate $z \in L_p(S \times T)$ by the definition

$$z(s, t) = [f(s)](t).$$

It is routine to verify that every element of $L_p(S \times T)$ arises in this way from a suitable $f \in L_p(S, L_p(T))$, and that

$$\|z\| = \left\{ \int |z(s,t)|^p \, ds \times dt \right\}^{1/p} = \left\{ \int \|f(s)\|^p \, ds \right\}^{1/p} = \|f\|.$$

(The Fubini theorem is used in this last calculation). ∎

1.53 THEOREM. *Let S and T be σ-finite measure spaces. Then $L_\infty(S) \otimes_\lambda L_\infty(T)$ is a subspace of $L_\infty(S \times T)$. This subspace is usually proper.*

PROOF. Let $X = L_1(S)$ and $Y = L_1(T)$. Then $X^* = L_\infty(S)$ and $Y^* = L_\infty(T)$. Now by 1.22 and 1.25, we have

$$(7) \qquad X^* \otimes_\lambda Y^* = \overline{FR}(X, Y^*) \subset \mathcal{L}(X, Y^*) = (X \otimes_\gamma Y)^*$$

where $\overline{FR}$ denotes the closure of the set of finite-rank operators. Using 1.16, we have

$$L_\infty(S) \otimes_\lambda L_\infty(T) \subset [L_1(S) \otimes_\gamma L_1(T)]^* = L_1(S \times T)^* = L_\infty(S \times T).$$

The inclusion in (7) is usually a proper one. ∎

CHAPTER 2

PROXIMINALITY

Let K be a subset of a Banach space X. We say that K is **proximinal** in X if to each x in X there corresponds at least one point $k_0 \in K$ such that $\|x - k_0\| = \inf_{k \in K} \|x - k\|$. The point k_0 is then called a **best approximation to** x from K. Some questions about proximinality in tensor product spaces now suggest themselves. Let X and Y be Banach spaces and let H be a subspace in Y. Under what conditions can it be asserted that $X \hat{\otimes} H$ is proximinal in $X \otimes_\alpha Y$? In particular, is the proximinality of H in Y a sufficient condition? This question is addressed in the first half of the chapter. (The answer depends on the crossnorm α.) In the second half, the central question is whether $(G \hat{\otimes} Y) + (X \hat{\otimes} H)$ is proximinal in $X \otimes_\alpha Y$, it being assumed that G and H are subspaces of X and Y respectively.

If K is proximinal in X, then any map which associates with each element of X one of its best approximations in K is called a **proximity map**. If K is such that each element of X has *exactly* one best approximation in K, then K is said to have the **Chebyshev** property.

If X is the space $C(S)$ and if α is taken to be the injective tensor norm λ, then an easy theorem is available.

2.1 THEOREM. *Let S be a compact Hausdorff space, and let H be a subspace of the Banach space Y. If there is a continuous proximity map of Y onto H, then $C(S) \hat{\otimes} H$ is proximinal in $C(S) \otimes_\lambda Y$, and in fact it has a continuous proximity map.*

PROOF. Let $A : Y \twoheadrightarrow H$ be a continuous proximity map. Then the mapping $A' : C(S, Y) \to C(S, H)$ defined by $A'f = A \circ f$ is a proximity map. The continuity of A' follows from 11.8. If $g \in C(S, H)$ then

$$\|f(s) - A(f(s))\| \le \|f(s) - g(s)\|$$

for all $s \in S$. Hence

$$\|f - A \circ f\| \le \|f - g\|.$$

Now use 1.14, which states that $C(S, Y) = C(S) \otimes_\lambda Y$, and 1.17, which states that

$$C(S) \otimes_\lambda H = C(S) \hat{\otimes} H. \quad \blacksquare$$

2.2 COROLLARY. *Let H be a subspace of a Banach space Y. If either*

(i) H is finite-dimensional and Chebyshev; or

(ii) Y is uniformly convex,

then $C(S, H)$ is proximinal in $C(S, Y)$.

PROOF. In both of these cases, a theorem from [**104**, p. 164] guarantees the continuity of the proximity map from Y onto H. The preceding theorem then applies. ■

An n-dimensional subspace H in $C(T)$ is called a **Haar** subspace if no function in $H\backslash 0$ has n zeros.

2.3 COROLLARY. *If H is a finite-dimensional Haar subspace in $C(T)$, then $C(S) \otimes_\lambda H$ is proximinal in $C(S \times T)$.*

PROOF. Every Haar subspace has the Chebyshev property [**104**, p. 114]. Hence by 2.2, $C(S) \otimes_\lambda H$ is proximinal in $C(S) \otimes_\lambda C(T)$. The latter is $C(S \times T)$ by 1.14. ■

2.4 THEOREM. *Let H be a closed subspace of the Banach space Y. Let S be a compact Hausdorff space. For each $f \in C(S, Y)$ we have*

$$\text{dist } (f, C(S, H)) = \text{dist } (f, \ell_\infty(S, H)) = \sup_s \text{dist } (f(s), H).$$

PROOF. Since $C(S, H) \subset \ell_\infty(S, H)$ it is clear that

$$\text{dist } (f, C(S, H)) \geq \text{dist } (f, \ell_\infty(S, H)).$$

If $g \in \ell_\infty(S, H)$ then for each $s \in S$,

$$\|f(s) - g(s)\| \geq \text{dist } (f(s), H).$$

It follows that

$$\|f - g\| \geq \sup_s \text{dist } (f(s), H)$$

and that

$$\text{dist } (f, \ell_\infty(S, H)) \geq \sup_s \text{dist } (f(s), H).$$

For the reverse inequality, let $\lambda > \sup_s \text{dist}(f(s), H.)$ For each $s \in S$ define

$$\Phi(s) = \{h \in H : \|f(s) - h\| \leq \lambda\}.$$

Then $\Phi(s)$ is a nonvoid, closed, convex subset of H. We shall prove that Φ is lower semicontinuous. Let $\mathcal{O}$ be an open set in H and put

$$\mathcal{O}^* = \{s \in S : \Phi(s) \cap \mathcal{O} \neq \square\}.$$

36

It is to be shown that O^* is open. Let $\sigma \in O^*$. Then $\Phi(\sigma) \cap O$ is nonvoid. Hence there exists an $h \in O$ such that $\|f(\sigma) - h\| \leq \lambda$. By the definition of λ, there exists an $h' \in H$ such that $\|f(\sigma) - h'\| < \lambda$. By convexity, there exists $h'' \in O$ such that $\|f(\sigma) - h''\| < \lambda$. Let $\mathcal{N}$ be a neighborhood of σ such that

$$\|f(s) - f(\sigma)\| < \lambda - \|f(\sigma) - h''\|$$

for all $s \in \mathcal{N}$. For any $s \in \mathcal{N}$ we then have

$$\|f(s) - h''\| \leq \|f(s) - f(\sigma)\| + \|f(\sigma) - h''\| < \lambda.$$

Hence $h'' \in \Phi(s) \cap O$, $s \in O^*$, $\mathcal{N} \subset O^*$, and O is open.

By the Michael Selection Theorem (11.14) there exists $g \in C(S, H)$ such that $g(s) \in \Phi(s)$ for all s. Hence $\|f(s) - g(s)\| \leq \lambda$ and $\|f - g\| \leq \lambda$. It follows that

$$\text{dist } (f, C(S, H)) \leq \lambda$$

and

$$\text{dist } (f, C(S, H)) \leq \sup_s \text{dist } (f(s), H). \quad \blacksquare$$

2.5 COROLLARY. Let H be a closed subspace of Y. In order that an element h of $C(S, H)$ be a best approximation to an element f of $C(S, Y)$ it is necessary and sufficient that for some point $s_0 \in S$, s_0 is a maximum point of $\|f(s) - h(s)\|$ and $h(s_0)$ is a best approximation to $f(s_0)$ from H.

2.6 THEOREM. Let S be an infinite compact metric space, and let T be a non-atomic measure space. If H is a finite-dimensional subspace of $L_1(T)$, then $C(S, H)$ is not proximinal in $C(S, L_1(T))$.

For a proof of this, as well as related results, see [**61**].

2.7 THEOREM. Let S and T be closed and bounded intervals of the real line. Then there exists a one-dimensional subspace H in $C(T)$ such that $C(S) \otimes H$ is not proximinal in $C(S \times T)$.

PROOF. The construction and proof are simpler if we use $S = [-1/2, 1/2]$ and $T = [-1, 1]$. Define

$$
\begin{aligned}
h(t) &= t && t \in T \\
f(s, t) &= 1 - |s - t| && (s, t) \in S \times T \\
F(s, t) &= f(s, t)/(1 - |s|) && (s, t) \in S \times T.
\end{aligned}
$$

37

One can show that the best approximation to f_s is h if $s > 0$ and $-h$ if $s < 0$. From this, dist $(f_s, H) = 1 - |s|$, where H denotes the subspace generated by h. It then follows that dist $(F_s, H) = 1$ for all s, and dist $(F, C(S) \otimes H) = 1$. But if $x \in \ell_\infty(S)$ and $\|F - x \otimes h\| \le 1$, then $x(s) = (1-s)^{-1}$ for $s > 0$ and $x(s) = -(1+s)^{-1}$ for $s < 0$. Thus x must be discontinuous at $s = 0$. The tedious details are omitted. ∎

2.8 THEOREM. *Let $(S, \mathcal{A}, \mu)$ be a finite, complete measure space. Let G be a finite-dimensional subspace in a Banach space X. Then $L_\infty(S, G)$ is proximinal in $L_\infty(S, X)$.*

PROOF. Let $f \in L_\infty(S, X)$. For each $s \in S$ define

$$\Phi(s) = \{g \in G : \|f(s) - g\| = \text{dist } (f(s), G)\}.$$

For each s, $\Phi(s)$ is a closed, bounded, and nonempty subset of G. We shall show that Φ is weakly measurable in the sense that for each compact subset K of G the set

$$K^* = \{s \in S : \Phi(s) \cap K \ne \square\}$$

is measurable in S. The set K^* can also be described as

$$K^* = \{s \in S : \inf_{g \in K} \|f(s) - g\| = \inf_{g \in G} \|f(s) - g\|\}.$$

Since $(S, \mathcal{A}, \mu)$ is complete, f is measurable in the classical sense, by 10.2. Since subtraction in X and the norm in X are continuous, the mapping $s \mapsto \|f(s) - g\|$ is measurable for each g. Hence the mapping $s \mapsto \inf_{g \in A} \|f(s) - g\|$ is measurable for any set A. It follows that K^* is measurable. By 11.17, there is a measurable function $h : S \to G$ such that $h(s) \in \Phi(s)$ for each $s \in S$. Since G is finite-dimensional, it is separable. Hence by 10.3, h is strongly measurable. Since $\|h(s)\| \le 2\|f(s)\|$, it follows that $\|h\| \le 2\|f\|$. Thus $h \in L_\infty(S, G)$. For any $k \in L_\infty(S, H)$ we have $\|f(s) - h(s)\| \le \|f(s) - k(s)\|$ for all s. Hence $\|f - h\| \le \|f - k\|$. This proves that h is a best approximation to f in $L_\infty(S, G)$. ∎

The previous theorem and the next do not fit into the tensor product framework, since the space $L_\infty(S, X)$ is generally not isometric to the space $L_\infty(S) \otimes_\lambda X$. For example, $\ell_\infty \otimes_\lambda \ell_1$ is a proper subspace of $\ell_\infty(\mathbb{N}, \ell_1)$. In order to see this, observe that by 1.26 $\ell_\infty \otimes_\lambda \ell_1$ is the subspace of compact operators in $\mathcal{L}(\ell_1, \ell_1)$. On the other hand, $\ell_\infty(\mathbb{N}, \ell_1)$ contains $\mathcal{L}(\ell_1, \ell_1)$.

2.9 THEOREM. *Let S be an arbitrary set, Y a Banach space, and H a proximinal subspace in Y. Then $\ell_\infty(S, H)$ is proximinal in $\ell_\infty(S, Y)$.*

PROOF. If $f \in \ell_\infty(S, Y)$ then one of its best approximations is the function g such that $\|f(s) - g(s)\| = \text{dist } (f(s), H)$ for each $s \in S$. ∎

38

2.10 LEMMA. Let $(S, \mathcal{A}, \mu)$ be a measure space, Y a Banach space, and H a subspace of Y. Then for each $f \in L_1(S, Y)$

$$\text{dist } (f, L_1(S, H)) = \int_S \text{dist } (f(s), H)ds.$$

PROOF. For any $g \in L_1(S, H)$ we have, by the definition in Chapter 10,

$$\|f - g\| = \int_S \|f(s) - g(s)\| ds \geq \int_S \text{dist } (f(s), H)ds.$$

By taking an infimum on g we obtain

$$\text{dist } (f, L_1(S, H)) \geq \int_S \text{dist } (f(s), H)ds.$$

For the reverse inequality, let $\epsilon > 0$, and let f' be a simple function in $L_1(S, Y)$ such that $\|f - f'\| < \epsilon$. Write $f' = \sum_{i=1}^n x_i y_i$ where the x_i are the characteristic functions of sets A_i in $\mathcal{A}$ and $y_i \in Y$. We may assume that $\sum_{i=1}^n x_i = 1$ and that $\mu(A_i) > 0$. Since $f' \in L_1(S, Y)$, we have $\|y_i\| \mu(A_i) < \infty$ for $1 \leq i \leq n$. If $\mu(A_i) < \infty$, select $h_i \in H$ so that

$$\|y_i - h_i\| < \text{dist } (y_i, H) + \frac{\epsilon}{n\mu(A_i)}.$$

If $\mu(A_i) = \infty$, put $h_i = 0$. Let $g = \sum_{i=1}^n x_i h_i$. Then $g \in L_1(S, H)$ and

$$\text{dist } (f, L_1(S, H)) \leq \epsilon + \text{dist } (f', L_1(S, H))$$
$$\leq \epsilon + \|f' - g\|$$
$$= \epsilon + \int \|f'(s) - g(s)\| ds$$
$$= \epsilon + \sum_{i=1}^n \int_{A_i} \|y_i - h_i\| ds$$
$$\leq \epsilon + \sum_{i=1}^n \|y_i - h_i\| \mu(A_i)$$
$$< \epsilon + \sum_{i=1}^n \left[\text{dist } (y_i, H)\mu(A_i) + \frac{\epsilon}{n} \right]$$
$$= 2\epsilon + \int \text{dist } (f'(s), H)ds$$
$$\leq 2\epsilon + \int [\text{dist } (f(s), H) + \|f(s) - f'(s)\|] ds$$
$$\leq 3\epsilon + \int \text{dist } (f(s), H)ds. \quad \blacksquare$$

2.11 COROLLARY. *Let H be a closed subspace of the Banach space Y. In order that an element g of $L_1(S, H)$ be a best approximation to an element f of $L_1(S, Y)$ it is necessary and sufficient that, for almost all $s \in S$, $g(s)$ be a best approximation in H to $f(s)$.*

2.12 LEMMA. *Let S be an arbitrary measure space, and let H be a proximinal subspace of a Banach space Y. Then each simple function f in $L_1(S, Y)$ has a best approximation w in $L_1(S, H)$ such that for all measurable sets A,*

$$\int_A \|w(s)\| ds \leq 2 \int_A \|f(s)\| ds.$$

PROOF. The simple function f can be written as $f = \sum_{i=1}^n x_i \otimes y_i$ where each x_i is the characteristic function of a measurable set A_i in S, $y_i \in Y$, and $\sum_{i=1}^n x_i = 1$. Let h_i be a best approximation to y_i in H, and set $w = \sum_{i=1}^n x_i \otimes h_i$. For any $v \in L_1(S, H)$ we have

$$\|f - v\| = \int \|f(s) - v(s)\| ds = \sum_{i=1}^n \int x_i(s) \|f(s) - v(s)\| ds$$

$$= \sum_{i=1}^n \int \|x_i(s) y_i - x_i(s) v(s)\| ds \geq \sum_{i=1}^n \int \|x_i(s) y_i - x_i(s) h_i\| ds$$

$$= \int \|\sum_{i=1}^n x_i(s) y_i - \sum_{i=1}^n x_i(s) h_i\| ds = \|f - w\|.$$

Observe also that for each s, $x_i(s)w$ is a best approximation to $x_i(s)f$. Hence $\|x_i(s)w\| \leq 2\|x_i(s)f\|$. When this inequality is integrated over an arbitrary set A, the inequality in the Lemma is obtained. ∎

2.13 THEOREM. *Let H be a reflexive subspace of the Banach space Y. Let S be a finite measure space. Then $L_1(S, H)$ is proximinal in $L_1(S, Y)$.*

PROOF. Let f be an element of $L_1(S, Y)$. Then there exists a sequence of simple functions f_n in $L_1(S, Y)$ converging to f. By 2.12, each f_n has a best approximation w_n in $L_1(S, H)$, and we may assume that

(1) $$\int_E \|w_n(s)\| ds \leq 2 \int_E \|f_n(s)\| ds$$

for all measurable $E \subset S$. The sequence $\{f, f_1, f_2, \ldots\}$ is norm-compact and hence weakly compact. It is therefore uniformly integrable by Theorem 4, page 104 of [55]. This means that to each $\epsilon > 0$ there corresponds a $\delta > 0$ such that

(2) $$\int_E \|f_n(s)\| ds < \epsilon \quad \text{whenever} \quad \mu(E) < \delta \quad \text{and} \quad n \in \mathbb{N}.$$

40

By inequalities (1) and (2), the sequence $\{w_n\}$ is uniformly integrable. It is also bounded. Hence by the Dunford compactness theorem [**55**, p. 101] the sequence $\{w_n\}$ has compact closure in the weak topology of $L_1(S, H)$. By taking subsequences and using the Eberlein theorem, we may assume that w_n converges weakly in $L_1(S, H)$ to an element w of $L_1(S, H)$. It therefore converges weakly in $L_1(S, Y)$ to w [**57**, p. 436, Problem 6]. By the weak lower-semicontinuity of the norm,

$$\|f - w\| \leq \liminf \|f_n - w_n\| = \liminf \operatorname{dist}\,(f_n, L_1(S, H)) = \operatorname{dist}\,(f, L_1(S, H)). \quad \blacksquare$$

2.14 LEMMA. *If P is a projection defined on a Banach space X, then the range and null space of P^* are proximinal subspaces in X^*.*

PROOF. It is elementary to prove that

$$P^*(X^*) = (\ker P)^\perp.$$

Since the annihilator of a subspace of X is weak*-closed in X^*, we conclude that the range of P^* is weak*-closed and proximinal. Now use the fact that $\ker P^* = (I - P^*)(X^*)$ to see that $\ker P^*$ is weak*-closed. $\quad \blacksquare$

2.15 LEMMA. *Let X and Y be Banach spaces. If P is a projection on X then the subspace*

$$M = \{AP : A \in \mathcal{L}(X, Y^*)\}$$

is complemented, weak*-closed, and proximinal in $\mathcal{L}(X, Y^*)$.

PROOF. Theorem 1.22 asserts that $\mathcal{L}(X, Y^*) = (X \otimes_\gamma Y)^*$. The space $\mathcal{L}(X, Y^*)$ has therefore a weak*-topology induced by its duality with $X \otimes_\gamma Y$. In this topology, a net A_α converges to 0 if and only if $(A_\alpha x)(y) \to 0$ for all $x \in X$ and $y \in Y$.

For $A \in \mathcal{L}(X, Y^*)$ define $p(A) = AP$. It is easily seen that p is a projection of $\mathcal{L}(X, Y^*)$ onto M. Now p is weak*-continuous because if $A_\alpha \to 0$ then $p(A_\alpha) \to 0$ since

$$(p(A_\alpha)x)(y) = (A_\alpha Px)(y) \to 0$$

for all (x, y). It follows that $I - p$ is weak*-continuous and that its null space, which is M, is weak*-closed. The proximinality of M follows from the fact that every weak*-closed set in a conjugate space is proximinal [**104**, p. 123]. $\quad \blacksquare$

In the same way one proves a dual result:

2.16 LEMMA. *Let X and Y be Banach spaces. Let Q be a projection on Y. Then the subspace*

$$N = \{Q^*A : A \in \mathcal{L}(X, Y^*)\}$$

is complemented, weak*-closed, and proximinal in $\mathcal{L}(X, Y^*)$.

2.17 THEOREM. *Let P and Q be projections on Banach spaces X and Y respectively. Then the subspace*

$$W = \{AP + Q^*B : A, B \in \mathcal{L}(X, Y^*)\}$$

is complemented, weak*-closed, and proximinal in $\mathcal{L}(X, Y^*)$.

PROOF. The subspace W is $M + N$, where M and N are defined in the two preceding lemmas. The projections p and q, defined by $p(A) = AP$ and $q(A) = Q^*A$, have ranges M and N. They commute with each other, since

$$p(q(A)) = (Q^*A)P = Q^*(AP) = q(p(A)).$$

Thus the Boolean sum $p \oplus q = p + q - pq$ is a projection of $\mathcal{L}(X, Y^*)$ onto W, by 11.1. Hence W is complemented and norm-closed. By 11.3, W is weak*-closed and proximinal. ∎

2.18 THEOREM. *Let G and H be finite-dimensional subspaces in conjugate Banach spaces X^* and Y^* respectively. Then $X^* \otimes H + G \otimes Y^*$ is complemented, weak*-closed, and proximinal in $\mathcal{L}(X, Y^*)$. It is therefore complemented and proximinal in $X^* \otimes_\lambda Y^*$.*

PROOF. We prove first that for an appropriate projection Q on Y, we have

$$X^* \otimes H = \{Q^*A : A \in \mathcal{L}(X, Y^*)\}.$$

Since H is finite-dimensional, there is a projection Q on Y such that $Q^*(Y^*) = H$. (Select a basis $h_1, \ldots, h_m$ for H and select y_i in Y so that $y_i(h_j) = \delta_{ij}$. Then let $Q = \sum_{i=1}^m h_i \otimes y_i$.) If N denotes the subspace of all Q^*A for $A \in \mathcal{L}(X, Y^*)$, then $X^* \otimes H \subset N$. Indeed, $Q^* \circ (\varphi \otimes h) = \varphi \otimes h$ for all $\varphi \otimes h \in X^* \otimes H$. For the reverse inclusion, note that each element of N belongs to $\mathcal{L}(X, H)$. Since H is finite-dimensional, $\mathcal{L}(X, H) = X^* \otimes H$.

A similar argument applies to show that

$$G \otimes Y^* = \{AP : A \in \mathcal{L}(X, Y^*)\}.$$

Then an application of 2.17 completes the proof. ∎

2.19 COROLLARY. *Let S and T be σ-finite measure spaces. If G and H are finite-dimensional subspaces in $L_\infty(S)$ and $L_\infty(T)$ respectively, then the subspace*

$$L_\infty(S) \otimes H + G \otimes L_\infty(T)$$

is weak*-closed, complemented, and proximinal in $L_\infty(S \times T)$.

PROOF. Let $X = L_1(S)$, $Y = L_1(T)$, and $W = X^* \otimes H + G \otimes Y^*$. Then $X^* = L_\infty(S)$ and $Y^* = L_\infty(T)$. By the preceding result, W is weak*-closed, complemented, and proximinal in $\mathcal{L}(X, Y^*)$. By 1.22 and 1.16,

$$\mathcal{L}(X, Y^*) = (X \otimes_\gamma Y)^* = (L_1(S) \otimes_\gamma L_1(T))^*$$
$$= L_1(S \times T)^* = L_\infty(S \times T). \quad \blacksquare$$

2.20 COROLLARY. *Let S and T be arbitrary sets, and let G and H be finite-dimensional subspaces in $\ell_\infty(S)$ and $\ell_\infty(T)$ respectively. Then*

$$\ell_\infty(S) \otimes H + G \otimes \ell_\infty(T)$$

is complemented, weak*-closed, and proximinal in $\ell_\infty(S \times T)$.

PROOF. This is a special case of the preceding result, taking counting measure on S and T. The σ-finiteness assumption is not needed, as $\ell_1(S)^* = \ell_\infty(S)$ for arbitrary sets [**103**, p. 129]. $\blacksquare$

2.21 THEOREM. *Let U and V be proximinal subspaces in a Banach space X. Assume that $U + V$ is closed, and that V has a proximity map A such that for each $c \in X$, the map $u \mapsto A(c - u)$ is weakly compact from U to V. Then $U + V$ is proximinal.*

PROOF. Let c be any element of X, and select $z_n \in U + V$ so that

$$\|c - z_n\| \to \text{dist}(c, U + V).$$

The sequence $\{z_n\}$ is bounded. Since $U + V$ is closed, 11.3 implies that z_n can be expressed as $u_n + v_n$, with $u_n \in U$, $v_n \in V$, and $\{u_n\}$ bounded. Put $v'_n = A(c - u_n)$. Since $\{u_n\}$ is bounded, our hypotheses imply that $\{v'_n\}$ lies in a weakly compact subset K of V.

By the Hahn-Banach theorem,

$$\text{dist}(x, U) = \sup\{\varphi(x) : \varphi \in U^\perp, \ \|\varphi\| = 1\}.$$

Hence the function $x \mapsto \text{dist}(x, U)$ is weakly lower semicontinuous, being the supremum of a family of weakly continuous functions [**148**, p. 39]. Let v be a point in K where $\text{dist}(c - v, U)$ is a minimum. Let u be a best approximation of $c - v$ in U. Then

$$\|c - v - u\| = \text{dist}(c - v, U) \leq \text{dist}(c - v'_n, U)$$
$$\leq \|c - v'_n - u_n\| \leq \|c - v_n - u_n\|.$$

43

By taking the limit, we get $\|c - v - u\| \leq \text{dist}\,(c, U + V)$. ∎

2.22 THEOREM. *Let G be a finite-dimensional subspace of $C(S)$ such that $G \otimes C(T)$ is proximinal in $C(S \times T)$. Let H be a finite-dimensional subspace of $C(T)$ having a Lipschitz proximity map. Then $C(S) \otimes H + G \otimes C(T)$ is proximinal in $C(S \times T)$.*

PROOF. By 2.1, the subspace $V = C(S) \otimes H$ is proximinal. Set $U = G \otimes C(T)$. By 11.2, $U + V$ is complemented and closed. Let A be a Lipschitz proximity map of $C(T)$ onto H, and define $(A'z)(s, t) = (Az_s)(t)$. Then A' is a proximity map of $C(S \times T)$ onto V. Define $\Gamma : U \to V$ by putting $\Gamma(u) = A'(z - u)$, where z is now fixed. By the following lemma, Γ is (weakly) compact. By 2.21, $U + V$ is proximinal. ∎

2.23 LEMMA. *Let G and H be finite-dimensional subspaces of $C(S)$ and $C(T)$ respectively. Let $U = G \otimes C(T)$ and $V = C(S) \otimes H$. Let $J : C(T) \to H$ be a Lipschitz map. Then for each $z \in C(S \times T)$ the map $\Gamma : U \to V$ defined by the following equation is compact:*

$$(\Gamma u)(s, t) = (J(z_s - u_s))(t).$$

PROOF. Let $B = \{u \in U : \|u\| \leq k\}$. We will show that the closure of $\Gamma(B)$ is compact in V. By the Ascoli Theorem, it suffices to prove that $\Gamma(B)$ is bounded and equicontinuous. Since J is a Lipschitz map, there is a constant λ such that

$$\|Jx - Jy\| \leq \lambda\|x - y\|.$$

Hence, for any $x \in C(T)$
$$\|Jx\| \leq \|Jx - J0\| + \|J0\|$$
$$\leq \lambda\|x\| + \|J0\|.$$

Thus $\Gamma(B)$ is bounded. The remainder of the proof concerns the equicontinuity of $\Gamma(B)$.

Select a basis $\{g_1, \ldots, g_n\}$ for G and functionals $\varphi_1, \ldots, \varphi_n \in C(S)^*$ so that $\|g_i\| = \|\phi_j\| = 1$ and $\phi_i(g_j) = \delta_{ij}$ (11.11). If $u \in B$, then $u = \sum_{i=1}^{n} g_i \otimes y_i$ and

$$|y_i(t)| = |\varphi_i(u^t)| \leq \|u^t\| \leq k.$$

Let (σ, τ) be a point of $S \times T$ at which the equicontinuiy of $\Gamma(B)$ is to be proved. Let $\epsilon > 0$. By the equicontinuity of the unit sphere in G there is a neighborhood $\mathcal{N}_1$ of σ such that $|g(s) - g(\sigma)| \leq \epsilon\|g\|$ for all $g \in G$ and all $s \in \mathcal{N}_1$. Similarly, there is a neighborhood $\mathcal{N}_2$ of τ such that $|h(t) - h(\tau)| \leq \epsilon\|h\|$ for all $h \in H$ and all $t \in \mathcal{N}_2$. By the equicontinuity of $\{z^t : t \in T\}$, (11.7), we can shrink the neighborhood $\mathcal{N}_1$ if necessary so that $|z^t(s) - z^t(\sigma)| < \epsilon$ for all $t \in T$ and $s \in \mathcal{N}_1$. Let $\mathcal{N} = \mathcal{N}_1 \times \mathcal{N}_2$, and let (s, t) be an arbitrary point of $\mathcal{N}$. Let u be an arbitrary point of B. Then, with $u = \sum_{i=1}^{n} g_i \otimes y_i$,

$$|(\Gamma u)(s,t) - (\Gamma u)(\sigma, \tau)|$$
$$\leq |(\Gamma u)(s,t) - (\Gamma u)(\sigma, t)| + |(\Gamma u)(\sigma, t) - (\Gamma u)(\sigma, \tau)|$$
$$= |J(z_s - u_s)(t) - J(z_\sigma - u_\sigma)(t)| + |J(z_\sigma - u_\sigma)(t) - J(z_\sigma - u_\sigma)(\tau)|$$
$$\leq \|J(z_s - u_s) - J(z_\sigma - u_\sigma)\| + \|J(z_\sigma - u_\sigma)\|\epsilon$$
$$\leq \lambda\|(z_s - u_s) - (z_\sigma - u_\sigma)\| + \{\lambda\|z_\sigma - u_\sigma\| + \|J0\|\}\epsilon$$
$$\leq \lambda\|z_s - z_\sigma\| + \lambda\|u_s - u_\sigma\| + \{\lambda\|z\| + \lambda\|u\| + \|J0\|\}\epsilon$$
$$\leq \lambda\epsilon + \lambda\|\sum_{i=1}^{n}[g_i(s) - g_i(\sigma)]y_i\| + \lambda\epsilon\|z\| + \lambda k\epsilon + \|J0\|\epsilon$$
$$\leq \lambda\epsilon + \lambda\epsilon n k + \lambda\epsilon\|z\| + \lambda k\epsilon + \|J0\|\epsilon.$$

Thus $\Gamma(B)$ is an equicontinuous subset of V. $\blacksquare$

2.24 THEOREM. *Let G and H be subspaces having linear proximity maps in Banach spaces X and Y respectively. For any uniform crossnorm α on $X \otimes Y$, the subspace*

$$X \,\bar{\otimes}\, H + G \,\bar{\otimes}\, Y$$

also has a linear proximity map, and is therefore proximinal and complemented in $X \otimes_\alpha Y$.

PROOF. Suppose that $P : X \twoheadrightarrow G$ and $Q : Y \twoheadrightarrow H$ are linear proximity maps. Then $P \otimes_\alpha I_Y$ and $I_X \otimes_\alpha Q$ are linear proximity maps of $X \otimes_\alpha Y$ onto $G \,\bar{\otimes}\, Y$ and $X \,\bar{\otimes}\, H$, respectively. One verifies readily that they commute with each other. Their Boolean sum is therefore a projection of $X \otimes_\alpha Y$ onto $X \,\bar{\otimes}\, H + G \,\bar{\otimes}\, Y$. In order to see that the Boolean sum of two commuting linear proximity maps (having a common domain) is again a linear proximity map, write

$$\|I - A - B + AB\| = \|(I - A)(I - B)\| \leq 1. \quad \blacksquare$$

2.25 EXAMPLE. Let $s_1, \ldots, s_n$ be points in a compact Hausdorff space S, and let G be the subspace (in fact an ideal) of $C(S)$ consisting of functions which vanish on $\{s_1, \ldots, s_n\}$. Similarly, let H be the ideal of functions in $C(T)$ which vanish on $\{t_1, \ldots, t_m\}$. Then $C(S) \otimes_\lambda H + G \otimes_\lambda C(T)$ has a linear proximity map. In order to employ the preceding theorem, we have to show that G and H have linear proximity maps. A linear proximity map for G is defined by

$$Px = x - \sum_{i=1}^{n} x(s_i)u_i$$

45

where $\{u_1, \ldots, u_n\}$ is a partition of unity in $C(S)$ satisfying $u_i(s_j) = \delta_{ij}$. For an arbitrary $g \in G$, the following inequality verifies that P is a proximity map:

$$\|x - Px\| = \|\Sigma x(s_i)u_i\| \leq \max_s \Sigma |x(s_i)|u_i(s)$$
$$\leq \max_i |x(s_i)| = \max_i |x(s_i) - g(s_i)| \leq \|x - g\|.$$

If $S = [a, b]$, one can take $a = s_1 < s_2 < \cdots < s_n = b$ and construct each u_i as a piecewise linear function (first-degree spline function) with knots at $s_1, \ldots, s_n$.

For further information on subspaces which have linear proximity maps, consult [50, 125].

2.26 THEOREM. *Let S and T be finite measure spaces. Let G and H be finite-dimensional subspaces in $L_1(S)$ and $L_1(T)$, respectively. Then $L_1(S) \otimes H + G \otimes L_1(T)$ is proximinal in $L_1(S \times T)$.*

PROOF. The subspaces $U = G \otimes L_1(T)$ and $V = L_1(S) \otimes H$ are proximinal in $L_1(S \times T)$, by 2.13 and 1.15. Note that $U + V$ is closed, by 11.2. Since V is proximinal, a proximity map $A : L_1(S \times T) \to V$ exists. By 2.11, A has the property that for almost all s, $(Az)_s$ is a best approximation to z_s in H. By 2.21, the proximinality of $U + V$ will follow if we can show that for each z the map $\Gamma(u) = A(z - u)$ is weakly compact from U to V.

Let $B = \{u \in U : \|u\| \leq k\}$. It is to be proved that $\Gamma(B)$ lies in a weakly compact subset of V. Select a biorthonormal system $\{h_i, \psi_i\}_1^m$ for H. Thus $\|h_i\| = \|\psi_i\| = 1$ and $y = \sum_{i=1}^m \psi_i(y)h_i$ for each $y \in H$. Now let $u \in B$ and $v = \Gamma(u)$. Then $v \in V$, and for almost all s, v_s is a best approximation to $z_s - u_s$. Hence

$$\|v_s\| \leq 2\|z_s - u_s\| \leq 2\|z_s\| + 2\|u_s\|.$$

Since $v_s \in H$, we have $v(s, t) = \sum_{i=1}^m \psi_i(v_s)h_i(t)$, whence

(3) $$|v(s, t)| \leq \|v_s\| \sum_{i=1}^m |h_i(t)| = \|v_s\|h(t) \leq 2(\|z_s\| + \|u_s\|)h(t).$$

Here we have put $h = \sum_{i=1}^m |h_i|$. With a similar argument involving a biorthonormal system for G we obtain

$$|u(s, t)| \leq \|u^t\|g(s) \qquad u \in U.$$

From the last inequality we obtain, for $u \in B$,

(4) $$\|u_s\| = \int_T |u(s, t)|dt \leq \int_T \|u^t\|g(s)dt = g(s)\|u\| \leq kg(s).$$

Thus, from (3) and (4), we have

(5) $$|v(s, t)| \leq \big[2\|z_s\| + 2kg(s)\big]h(t) \equiv F(s, t).$$

46

Since the function on the right is an element of $L_1(S \times T)$ and is independent of u, we have

$$\lim_{\sigma(E) \to 0} \iint_E |v(s,t)| \, ds \, dt \leq \lim_{\sigma(E) \to 0} \iint_E F(s,t) \, ds \, dt = 0$$

uniformly in v, as u ranges over B. By the Dunford Compactness Theorem, $\Gamma(B)$ lies in a weakly compact set. Note that the boundedness of $\Gamma(B)$ is implied by (5). ∎

The following example, due to von Golitschek, shows how the approximation problems considered in this chapter may be sensitive to alterations in the domain.

2.27 EXAMPLE. Let $S = T = [0, 0.3]$, and define the domain

$$D = \{(s,t) \ : \ 0 \leq s/2 \leq t \leq s \leq 0.3\}.$$

Let W be the subspace of $C(S \times T)$ consisting of functions

$$w(s,t) = x(s) + y(t) \qquad x \in C(S), \ y \in C(T).$$

By 2.22, W is proximinal in $C(S \times T)$. But if R denotes the restriction map from $C(S \times T)$ to $C(D)$, then $R(W)$ is *not* proximinal in $C(D)$; in fact, it is not closed. This can be seen by considering the following functions in W :

$$w_n(s,t) = x_n(s) + y_n(t) \qquad (n = 3, 4, \ldots)$$

where $y_n(t) = -x_n(t)$ and

$$x_n(s) = \begin{cases} \log \log 1/s & \text{if } 1/n \leq s \leq 0.3 \\ \log \log n & \text{if } 0 \leq s \leq 1/n. \end{cases}$$

One can prove the following facts.

(i) The sequence $\|w_n\|_D$ is bounded.

(ii) Every representation of Rw_n has the form $(x_n - c_n) + (y_n + c_n)$ for some constant c_n. (The topological nature of D must be taken into account in proving this equation.)

(iii) For any $c_n \in \mathbb{R}$,

$$\|x_n - c_n\| + \|y_n + c_n\| \to \infty.$$

By 11.2, $R(W)$ is not closed.

CHAPTER 3

THE ALTERNATING ALGORITHM

Let X be a Banach space and U a subspace of X. Recall that a mapping $A : X \twoheadrightarrow U$ is a **proximity** map if

$$\text{dist}\,(x, U) = \|x - Ax\| \qquad (x \in X).$$

Thus, Ax is a best approximation to x in U. It is clear that U is proximinal if and only if it has a proximity map. Some subspaces have continuous or even linear proximity maps, but these properties are exceptional. In Hilbert space, the orthogonal projection onto a closed subspace is its proximity map.

Now suppose that two subspaces with proximity maps are given, say $A : X \twoheadrightarrow U$ and $B : X \twoheadrightarrow V$. Is it possible to construct from A and B a proximity map for $\overline{U + V}$? It is natural to try the following iterative process. Starting with x_0, we compute $x_1 = x_0 - Bx_0$, $x_2 = x_1 - Ax_1$, $x_3 = x_2 - Bx_2$, and so forth. It is clear that $\|x_0\| \geq \|x_1\| \geq \|x_2\| \geq \cdots$ and that each x_n differs from x_0 by an element of $U + V$. The procedure just described is called the **Alternating Algorithm**. It originated with von Neumann in 1933 [**135**]. When the circumstances are favorable, $\lim_{n\to\infty} x_n$ exists, and $x_0 - \lim_{n\to\infty} x_n$ is a best approximation of x_0 in $\overline{U + V}$.

A succinct description of the algorithm goes as follows. Let $E = (I - A)(I - B)$. Then $x_{2n} = E^n x_0$. Observe that if the proximity maps A and B are linear, then $E = I - (A \oplus B)$. In this case, A and B are (bounded, linear) projections, and the theory of Boolean sums (11.1) can be applied. Thus we have:

3.1 THEOREM. *If the proximity maps A and B are linear and if $ABA = BA$, then the alternating algorithm produces a solution in two steps; i.e., $x_0 - x_2$ is a best approximation of x_0 in $U + V$.*

A basic result, established by von Neumann, is that in Hilbert space the alternating algorithm is effective for any pair of closed subspaces. This theorem is proved below (Theorem 3.8).

A **smooth point** in a Banach space X is a point x such that there exists a unique functional $\varphi \in X^*$ satisfying $\|\varphi\| = 1$ and $\varphi(x) = \|x\|$. If every nonzero point of X is a smooth point, then X is said to be **smooth**.

3.2 LEMMA. Let $A : X \twoheadrightarrow U$ and $B : X \twoheadrightarrow V$ be proximity maps on a Banach space X, and let $E = (I - A)(I - B)$. Consider the following properties of a point x in X:

(i) x is a fixed point of E; i.e., $Ex = x$

(ii) $\|x\| = $ dist $(x, U) = $ dist (x, V)

(iii) $\|x\| = $ dist $(x, U + V)$.

Then (i) $\Rightarrow$ (ii) always; (iii) $\Rightarrow$ (ii) always; (ii) $\Rightarrow$ (i) if U and V are Chebyshev subspaces; and (ii) $\Rightarrow$ (iii) if x is a smooth point.

PROOF. The equation $x = Ex$ reduces to $x = x - Bx - A(x - Bx)$ and to $-Bx = A(x - Bx)$. This shows that $Bx \in U$. Therefore

$$\|x\| = \|x - Bx - A(x - Bx)\| = \text{dist } (x - Bx, U) = \text{dist } (x, U).$$

Then it follows that

$$\|x\| = \text{dist } (x - Bx, U) \leq \|x - Bx\| = \text{dist } (x, V) \leq \|x\|.$$

This proves that (i) $\Rightarrow$ (ii). If (ii) is true, then by Singer's Theorem [159, p. 18], there exist functionals $\phi \in U^\perp$ and $\psi \in V^\perp$ such that $\|\phi\| = \|\psi\| = 1$ and $\|x\| = \phi(x) = \psi(x)$. If x is a smooth point, then $\phi = \psi$, and consequently $\phi \in U^\perp \cap V^\perp = (U + V)^\perp$. By Singer's Theorem, $\|x\| = $ dist $(x, U + V)$. Thus (ii) $\Rightarrow$ (iii) if x is a smooth point. The implication (iii) $\Rightarrow$ (ii) is trivial: $\|x\| = $ dist $(x, U + V) \leq $ dist $(x, U) \leq \|x\|$. Similarly, $\|x\| = $ dist (x, V). If (ii) is true, then 0 is a best approximation of x in U. If U is a Chebyshev subspace, then $Ax = 0$. Similarly $Bx = 0$ if V is a Chebyshev subspace. Then we see at once that $Ex = x$. ∎

3.3 THEOREM. If X is a smooth space, and if E is contractive, then the sequence $\{x_{2n}\}$ defined by the alternating algorithm is convergent. Furthermore, $x_0 - \lim x_{2n}$ is a best approximation of x_0 in $\overline{U + V}$.

PROOF. By Banach's Contractive Mapping Theorem, the map E has a unique fixed point, x^*, and the sequence $x_{2n} = E^n x_0$ converges to it. By 3.2, $\|x^*\| = $ dist $(x^*, U + V)$. By induction we see easily that $x_0 - x_n \in U + V$ for all n. Hence $x_0 - x^* \in \overline{U + V}$. Then

$$\|x_0 - (x_0 - x^*)\| = \|x^*\| = \text{dist } (x^*, U + V)$$
$$= \text{dist } (x^*, \overline{U + V}) = \text{dist } (x^* + (x_0 - x^*), \overline{U + V})$$
$$= \text{dist } (x_0, \overline{U + V}). \quad \blacksquare$$

The **modulus of convexity** of a Banach space X is the function $\delta : (0, 2] \to [0, 1]$ defined by

$$\delta(\epsilon) = \inf \left\{ 1 - \frac{1}{2}\|x + y\| : \|x\| \leq 1, \|y\| \leq 1, \|x - y\| \geq \epsilon \right\}.$$

3.4 LEMMA. *If* $0 < \|x\| \leq \frac{1}{2}\|x + y\|$ *then*

$$\|y\|\,\delta(\|x - y\|/\|y\|) \leq \|y\| - \|x\|.$$

PROOF. We have

$$0 < \|x\| \leq \frac{1}{2}\|x + y\| \leq \frac{1}{2}\|x\| + \frac{1}{2}\|y\|,$$

so $\|x\| \leq \|y\|$. Let $u = x/\|y\|$, $v = y/\|y\|$, and $\epsilon = \|u - v\|$. Then by the definition of δ,

$$\delta(\epsilon) \leq 1 - \frac{1}{2}\|u + v\|.$$

When we substitute the definitions of ϵ, u and v the result is

$$\|y\|\,\delta(\|x - y\|/\|y\|) \leq \|y\| - \frac{1}{2}\|x + y\| \leq \|y\| - \|x\|. \qquad \blacksquare$$

A Banach space is said to be **uniformly convex** if its modulus of convexity is a strictly positive function.

3.5 LEMMA. *If X is uniformly convex, then the iterates in the alternating algorithm satisfy* $\|x_n - x_{n+1}\| \to 0$.

PROOF. The conclusion is obvious if $\|x_n\| \to 0$. We assume therefore that

$$\lim_{n \to \infty} \|x_n\| = k > 0.$$

For appropriate points w_n in $U \cup V$, we have

$$\begin{aligned}
\|x_{n+1}\| = \|x_n - w_n\| &\leq \|x_n - \frac{1}{2}w_n\| \\
&= \frac{1}{2}\|x_n + (x_n - w_n)\| \\
&= \frac{1}{2}\|x_n + x_{n+1}\|.
\end{aligned}$$

By 3.4, it follows that

$$\|x_n\|\,\delta(\|x_n - x_{n+1}\|/\|x_n\| \leq \|x_n\| - \|x_{n+1}\| \longrightarrow 0.$$

Since $\|x_n\| \geq k$, we have

$$\delta(\|x_n - x_{n+1}\|/\|x_n\|) \longrightarrow 0.$$

By the uniform convexity, $\|x_n - x_{n+1}\| \to 0$. $\blacksquare$

A mapping F defined on a subset of a normed space is said to be **nonexpansive** if

$$\|Fx - Fy\| \leq \|x - y\|.$$

It is said to be **odd** if $F(-x) = -Fx$.

The following theorem is due to Baillon, Bruck, and Reich [**19**].

3.6 THEOREM. *Let C be a symmetric, closed, convex set in a uniformly convex Banach space. Let $T : C \to C$ be a map which is odd and nonexpansive. If ξ is a point in C for which $\lim_n(T^n\xi - T^{n+1}\xi) = 0$, then the sequence $\{T^n\xi\}$ converges to a fixed point of T.*

PROOF. Since T is odd, $T(0) = 0$. Therefore,

$$\|T^n\xi\| = \|T^n\xi - T0\| \leq \|T^{n-1}\xi - 0\| = \|T^{n-1}\xi\|.$$

It follows that the limit $\lambda = \lim_n\|T^n\xi\|$ exists. We assume that $\lambda > 0$ because the other case is trivial. For each i,

$$\lim_n(T^n\xi - T^{n+i}\xi) = \lim[(T^n\xi - T^{n+1}\xi) + \cdots + (T^{n+i-1}\xi - T^{n+i}\xi)] = 0.$$

Furthermore,

$$\|T^n\xi + T^{n+i}\xi\| = \|T^n\xi - T^{n+i}(-\xi)\| \leq \|T^{n-1}\xi - T^{n+i-1}(-\xi)\|$$
$$= \|T^{n-1}\xi + T^{n+i-1}\xi\|.$$

We have now

$$\lambda \leq \|T^n\xi\| \leq \frac{1}{2}\|T^n\xi + T^{n+i}\xi\| + \frac{1}{2}\|T^n\xi - T^{n+i}\xi\|.$$

By taking the limit as $n \to \infty$, we get

$$\lambda \leq \lim_n \frac{1}{2}\|T^n\xi + T^{n+i}\xi\|.$$

Because of the monotonicity established previously,

$$\lambda \leq \frac{1}{2}\|T^n\xi + T^m\xi\| \qquad \text{(all } n \text{ and } m\text{)}.$$

Let $\epsilon \mapsto \delta(\epsilon)$ be the modulus of uniform convexity. We will show that the sequence $\{T^n\xi\}$ has the Cauchy property. Let $\epsilon > 0$. Select θ so that $0 < \theta < 1$ and $\lambda/(\lambda + \theta) > 1 - \delta(\epsilon/(\lambda + 1))$. Select an integer n such that $\|T^n\xi\| \leq \lambda + \theta$. Let $i, j \geq n$.

Then

$$\left\|\frac{T^i\xi}{\lambda + \theta}\right\| \leq 1, \qquad \left\|\frac{T^j\xi}{\lambda + \theta}\right\| \leq 1,$$

$$\frac{1}{2}\left\|\frac{T^i\xi}{\lambda + \theta} + \frac{T^j\xi}{\lambda + \theta}\right\| \geq \frac{\lambda}{\lambda + \theta} > 1 - \delta\left(\frac{\epsilon}{\lambda + 1}\right).$$

51

From the definition of the modulus of convexity, it follows that

$$\left\| \frac{T^i \xi}{\lambda + \theta} - \frac{T^j \xi}{\lambda + \theta} \right\| < \frac{\epsilon}{\lambda + 1} < \frac{\epsilon}{\lambda + \theta}.$$

Hence $\|T^i \xi - T^j \xi\| < \epsilon$, and the Cauchy property is established. Since the space is complete, the limit $z = \lim_n T^n \xi$ exists. Since C is closed, $z \in C$. Since T is continuous, $Tz = T(\lim_n T^n \xi) = \lim T^{n+1} \xi = z.$ ∎

3.7 THEOREM. *If X is smooth and uniformly convex, and if*

$$A : X \twoheadrightarrow U \quad \text{and} \quad B : X \twoheadrightarrow V$$

are proximity maps such that $I - A$ and $I - B$ are nonexpansive, then the alternating algorithm is effective. Thus $\lim_{n \to \infty} (I - E^n)$ is the proximity map for $\overline{U + V}$.

PROOF. It is easy to see that the map $E = (I - A)(I - B)$ satisfies the hypotheses in 3.6. By 3.6, $\{E^n x\}$ converges to a fixed point of E. By 3.2, $x - \lim E^n x$ is a best approximation of x in $\overline{U + V}$. ∎

3.8 THEOREM. *In Hilbert space, the alternating algorithm is effective. Thus*

$$\lim_{n \to \infty} (I - E^n)$$

is the orthogonal projection onto $\overline{U + V}$.

PROOF. The proximity maps $A : X \twoheadrightarrow U$ and $B : X \twoheadrightarrow V$ are linear and of norm 1 in this case. Hence $I - A$ and $I - B$ are linear and of norm 1. Now apply 3.7. ∎

3.9 LEMMA. *Let U and V be proximinal subspaces in a Banach space X, and let $\{x_n\}$ be a sequence produced by the alternating algorithm starting at x. Assume that $\|x_n\| \geq k > 0$ for all n. If n is even select $\varphi_n \in U^\perp$ and if n is odd select $\varphi_n \in V^\perp$ so that*

$$\|\varphi_n\| = \|x_n\| \quad \text{and} \quad \varphi_n(x_n) = \|x_n\|^2.$$

Then

$$\delta^* (\|\varphi_n - \varphi_{n+1}\| / \|\varphi_n\|) \longrightarrow 0,$$

where δ^* is the modulus of convexity of $U^\perp + V^\perp$.

PROOF.

$$\begin{aligned}
\frac{1}{2}\|\varphi_{2n+1} + \varphi_{2n}\| &\geq \frac{1}{2}(\varphi_{2n+1} + \varphi_{2n})(x_{2n})/\|x_{2n}\| \\
&= \{\varphi_{2n+1}(x_{2n}) + \|x_{2n}\|^2\}/2\|x_{2n}\| \\
&= \{\varphi_{2n+1}(x_{2n+1}) + \|x_{2n}\|^2\}/2\|x_{2n}\| \\
&= \{\|x_{2n+1}\|^2 + \|x_{2n}\|^2\}/2\|x_{2n}\| \\
&\geq \|x_{2n+1}\| = \|\varphi_{2n+1}\| > 0.
\end{aligned}$$

52

By 3.4,

$$\|\varphi_{2n}\| \delta^* (\|\varphi_{2n+1} - \varphi_{2n}\| / \|\varphi_{2n}\|) \le \|\varphi_{2n}\| - \|\varphi_{2n+1}\|.$$

A similar result holds true for $2n$ and $2n - 1$. Hence for all n,

$$\|\varphi_n\| \delta^* (\|\varphi_{n+1} - \varphi_n\| / \|\varphi_n\|) \le \|\varphi_n\| - \|\varphi_{n+1}\|$$

and

$$\delta^* (\|\varphi_{n+1} - \varphi_n\| / \|\varphi_n\|) \le \|x_n\| - \|x_{n+1}\| \longrightarrow 0. \quad \blacksquare$$

3.10 THEOREM. *Let U, V and $U + V$ be closed subspaces in a Banach space X. If X and $U^\perp + V^\perp$ are uniformly convex then the alternating method produces best approximations in $U + V$.*

PROOF. Let $\{x_n\}$ be the sequence produced by the alternating algorithm, starting with x. For each n, there is an element $\varphi_n \in X^*$ such that $\|\varphi_n\| = \|x_n\|$ and $\varphi_n(x_n) = \|x_n\|^2$. Since $x_{2k} = x_{2k-1} - Ax_{2k-1}$, and $x_{2k+1} = x_{2k} - Bx_{2k}$, we can assume further that $\varphi_{2k} \in U^\perp$ and $\varphi_{2k+1} \in V^\perp$.

Henceforth, w will denote an arbitrary element of W satisfying $\|x - w\| \le \|x\|$. Since $U + V$ is closed, there exists (by 11.3) a constant c such that w has a representation $w = u + v$, with $u \in U$, $v \in V$, and $\|u\| + \|v\| \le c\|w\| \le 2c\|x\|$.

Select $\bar{u}_n \in U$ and $\bar{v}_n \in V$ so that $\bar{u}_n + \bar{v}_n = x - x_{2n}$ and $\|\bar{u}_n\| + \|\bar{v}_n\| \le c\|x - x_{2n}\| \le 2c\|x\|$. Then

$$
\begin{aligned}
\|x_{2n}\|^2 &= \varphi_{2n}(x_{2n}) = \varphi_{2n}(x - \bar{u}_n - \bar{v}_n) = \varphi_{2n}(x - \bar{v}_n) \\
&= \varphi_{2n}(x - w) + \varphi_{2n}(w - \bar{v}_n) \\
&\le \|\varphi_{2n}\| \; \|x - w\| + \varphi_{2n}(u + v - \bar{v}_n) \\
&= \|x_{2n}\| \; \|x - w\| + \varphi_{2n}(v - \bar{v}_n) \\
&= \|x_{2n}\| \; \|x - w\| + (\varphi_{2n} - \varphi_{2n-1})(v - \bar{v}_n) \\
&\le \|x_{2n}\| \; \|x - w\| + \|\varphi_{2n} - \varphi_{2n-1}\| \, \{\|v\| + \|\bar{v}_n\|\} \\
&\le \|x_{2n}\| \; \|x - w\| + 4c\|x\| \; \|\varphi_{2n} - \varphi_{2n-1}\|.
\end{aligned}
$$

In this inequality, we now take an infimum on w, getting

$$\|x_{2n}\|^2 \le \rho\|x_{2n}\| + 4c\|x\| \; \|\varphi_{2n} - \varphi_{2n-1}\|$$

where $\rho = \text{dist}(x, U + V)$. If $\lim_n \|x_n\| = 0$, we are finished. In the other case, 3.9 shows that $\|\varphi_n - \varphi_{n-1}\| \to 0$, and so we have $\lim \|x_{2n}\| = \rho$. Hence $\lim \|x_{2n+1}\| = \rho$. This means that the points $w_n = x - x_n$ form a minimizing sequence in $U + V$ for approximating x. Since X is uniformly convex, $\lim w_n$ exists and is a best approximation of x. $\quad \blacksquare$

The next few lemmas are needed to establish Theorem 3.16 concerning geometric convergence of the alternating algorithm.

3.11 LEMMA. *If A and B are linear proximity maps and the limit $x^* = \lim_{n \to \infty} x_n$ exists in the alternating algorithm, then $\|x_{2n+2} - x^*\| \leq \theta \|x_{2n} - x^*\|$, with $\theta = \|E|W\|$.*

PROOF. Since x^* is a fixed point of E, and E is linear,

$$\|x_{2n+2} - x^*\| = \|Ex_{2n} - Ex^*\| = \|E(x_{2n} - x^*\|$$
$$\leq \|E|W\| \; \|x_{2n} - x^*\|.$$

Here it is also necessary to note that $x_{2n} - x^* \in W = \overline{U + V}$. ∎

3.12 DEFINITION. *If U and V are subspaces of a normed space X, we define their inclination to be*

$$\mathrm{incl}\,(U, V) = \sup\{\phi(v) : \phi \in X^*, \; u \in U, \; v \in V, \; \|\phi\| = \|u\| = \|v\| = \phi(u) = 1\}.$$

3.13 LEMMA. *If $U \cap V = 0$ and if $U + V$ is closed and uniformly convex, then $\mathrm{incl}\,(U, V) < 1$.*

PROOF. Assume the hypotheses and that $\mathrm{incl}\,(U, V) = 1$. Then there exist

$$\phi_n \in X^*, \; u_n \in U, \quad \text{and} \quad v_n \in V$$

such that $\|\phi_n\| = \|u_n\| = \|v_n\| = \phi_n(u_n) = 1$ and $\phi_n(v_n) \to 1$. Since

$$1 + \phi_n(v_n) = \phi_n(u_n + v_n) \leq \|u_n + v_n\| \leq 2$$

and since $\phi_n(v_n) \to 1$, we have $\|u_n + v_n\| \to 2$. Since $U + V$ is uniformly convex,

$$\|u_n - v_n\| \to 0.$$

Since $U \cap V = 0$, each element of $U + V$ has a unique representation as a sum of an element of U and an element of V. Since $U + V$ is closed, there exists by 11.3 a constant c such that $\|u_n\| + \|v_n\| \leq c\|u_n - v_n\|$. This is not possible, as $\|u_n\| = \|v_n\| = 1$ and $\|u_n - v_n\| \to 0$. ∎

A subspace U in a Banach space X is said to be **smooth** if each nonzero element u of U is a smooth point of X.

3.14 LEMMA. *If P is a norm-1 projection of a Banach space X onto a smooth subspace U, then for any other subspace V,*

$$\mathrm{incl}\,(U, V) = \|P|V\|.$$

PROOF. Since U is smooth, there corresponds to each $u \in U \backslash 0$ a unique functional $\phi_u \in X^*$ such that $\|\phi_u\| = 1$ and $\phi_u(u) = \|u\|$. By the Hahn-Banach Theorem, there corresponds to each $x \in X$ a functional $\psi_x \in X^*$ such that $\|\psi_x\| = 1$, $\psi_x \perp \ker P$ and $\psi_x(x) = \text{dist}(x, \ker P)$. Since $\|P\| = 1$, $I - P$ is a proximity map. Its range is of course $\ker P$. Hence $\psi_x(x) = \|x - (I - P)x\| = \|Px\|$. If $u \in U$ then $\psi_u(u) = \|u\|$, and so $\psi_u = \phi_u$. Hence $\phi_u \in (\ker P)^\perp$. This establishes that $\phi_u(x - Px) = 0$ for all $u \in U$ and all $x \in X$. Now

$$\|P|V\| = \sup_{\substack{v \in V \\ \|v\| = 1}} \|Pv\| = \sup_{\substack{v \in V \\ \|v\| = 1}} \sup_{\substack{u \in U \\ \|u\| = 1}} \phi_u(Pv) = \sup_{\substack{v \in V \\ \|v\| = 1}} \sup_{\substack{u \in U \\ \|u\| = 1}} \phi_u(v).$$

The latter expression is $\text{incl}(U, V)$ since U is smooth. ∎

3.15 LEMMA. *Suppose that, in the alternating algorithm, A and B are linear, while $W (= U + V)$ is smooth and uniformly convex. If $(\ker A \cap W) + (\ker B \cap W)$ is closed, then $\|E|W\| < 1$.*

PROOF. First we prove that $\ker A \cap \ker B \cap W = 0$. If w is an element of this set, then $Aw = Bw = 0$. Consequently $\|w\| = \text{dist}(w, U) = \text{dist}(w, V)$. By 3.2, $\|w\| = \text{dist}(w, U + V) = 0$. Now by 3.13 and 3.14 we have

$$\|E|W\| = \sup_{\substack{w \in W \\ \|w\| = 1}} \|(I - A)(I - B)w\| \leq \sup_{\substack{x \in W \cap \ker B \\ \|x\| \leq 1}} \|(I - A)x\|$$

$$= \text{incl}\,[\ker A \cap W, \ker B \cap W] < 1. \quad ∎$$

3.16 THEOREM. *In the alternating algorithm, assume that the proximity maps A and B are linear, while $W (= U + V)$ is smooth and uniformly convex. If $(\ker A \cap W) + (\ker B \cap W)$ is closed, then the iterates $x - x_n$ converge geometrically to a best approximation of x in $\overline{W}$.*

PROOF. Use 3.7, 3.11, and 3.15. ∎

55

CHAPTER 4

CENTRAL PROXIMITY MAPS

A proximity map A from a Banach space X onto a subspace U is said to be a *central proximity map* if for all $x \in X$ and all $u \in U$,

$$\|x - Ax + u\| = \|x - Ax - u\|.$$

Such maps are rare, but there are notable examples. The concept is due to Golomb [78]. He observed that the analysis of Diliberto and Straus depended only on the centrality property of the proximity maps. The Diliberto-Straus-Golomb theory is presented next, in a form modeled on that in [78]. We refer to the definition of the alternating algorithm as given at the beginning of Chapter 3.

4.1 LEMMA. *If the proximity maps A and B are central, then the iterates in the alternating algorithm have these properties:*

(i) $\|x_n\| = \|x_n - 2x_{n+1}\|$

(ii) *If $\phi \in X^*$ and $\|\phi\| \leq 1$, then $\phi(x_n) \geq 2\phi(x_{n+1}) - \|x_n\|$.*

PROOF. If n is odd, then (because A is a central proximity map)

$$\|2x_{n+1} - x_n\| = \|x_{n+1} + (x_{n+1} - x_n)\| = \|x_n - Ax_n - Ax_n\| = \|x_n - Ax_n + Ax_n\| = \|x_n\|.$$

The proof for even n is the same, except that B replaces A. Now if $\|\phi\| \leq 1$, then

$$\phi(2x_{n+1} - x_n) \leq \|2x_{n+1} - x_n\| = \|x_n\|.$$

This yields (ii) immediately. ∎

4.2 LEMMA. *Let $A : X \longrightarrow U$ be a central proximity map. Corresponding to each $x \in X$ and $\phi \in X^*$ there is an element $\psi \in X^*$ such that $\|\psi\| = \|\phi\|$, $\phi + \psi \in U^\perp$, and $\psi(x - Ax) = \phi(x - Ax)$.*

PROOF. Define ψ on the subspace M generated by U and $x - Ax$ by putting

$$\psi\big(u + \lambda(x - Ax)\big) = \phi\big(-u + \lambda(x - Ax)\big).$$

Here, $u \in U$ and $\lambda \in \mathbf{R}$. It is clear that $(\phi + \psi)(u) = 0$ and that $\phi(x - Ax) = \psi(x - Ax)$. Since A is a central proximity map, we have the following equation (in which $\lambda \neq 0$ and $u' = u/\lambda$)

$$|\psi(u + \lambda(x - Ax))| = |\phi(-u + \lambda(x - Ax))| = |\lambda|\,|\phi(-u' + x - Ax)|$$

$$\leq |\lambda|\,\|\phi\|\,\| - u' + x - Ax\| = |\lambda|\,\|\phi\|\,\|u' + x - Ax\| = \|\phi\|\,\|u + \lambda(x - Ax)\|.$$

Thus the norm of ψ on M does not exceed $\|\phi\|$. An application of the Hahn-Banach Theorem completes the proof. ∎

4.3 THEOREM. *Let U and V be subspaces of a Banach space having central proximity maps A and B, respectively. If $U+V$ is closed, then the sequence $\{x_n\}$ generated by the alternating algorithm has the property $\|x_n\| \downarrow \operatorname{dist}(x, U + V)$.*

PROOF. Fix two even integers m and k. Let $n = m + k$. By the Hahn-Banach theorem there exists a functional $\phi_0 \in U^\perp$ such that $\|\phi_0\| = 1$ and $\phi_0(x_n) = \|x_n\|$. By 4.2, we can define inductively the functionals $\phi_1, \phi_2, \ldots, \phi_k$ so that

(1) $\phi_i + \phi_{i-1} \in U^\perp$ if i is odd, $1 \leq i \leq k$

(2) $\phi_i + \phi_{i-1} \in V^\perp$ if i is even, $1 \leq i \leq k$

(3) $\|\phi_i\| \leq 1$, $0 \leq i \leq k$

(4) $\phi_i(x_{n-i+1}) = \phi_{i-1}(x_{n-i+1})$, $1 \leq i \leq k$.

We now assert that for each r in the range $1 \leq r \leq k$,

(5) $\phi_i(x_{n-r}) \geq \|x_m\| - 2^r(\|x_m\| - \|x_n\|)$, $0 \leq i \leq r$.

This will be proved by induction on r. For $r = 1$ the proof is contained in the following inequality, based upon the fact that ϕ_0 and ϕ_1 belong to $U^\perp$:

$$\phi_1(x_{n-1}) = \phi_1(x_n + Ax_{n-1}) = \phi_1(x_n) = \phi_0(x_n) = \phi_0(x_{n-1})$$

$$= \|x_n\| \geq \|x_n\| + (\|x_n\| - \|x_m\|) = \|x_m\| - 2(\|x_m\| - \|x_n\|).$$

We now assume the validity of (5) for $r = t < k$, and prove its validity for $r = t + 1$. By 4.1,

$$\phi_{t+1}(x_{n-t-1}) \geq 2\phi_{t+1}(x_{n-t}) - \|x_{n-t-1}\|$$

$$\geq 2\phi_t(x_{n-t}) - \|x_m\|$$

$$\geq 2\{\|x_m\| - 2^t(\|x_m\| - \|x_n\|)\} - \|x_m\|$$

$$= \|x_m\| - 2^{t+1}(\|x_m\| - \|x_n\|).$$

Thus, the inequality (5) is true for $r = t + 1 = i$. If $0 \leq i \leq t$, then as before,

$$\phi_i(x_{n-t-1}) \geq 2\phi_i(x_{n-t}) - \|x_{n-t-1}\|$$

$$\geq 2\{\|x_m\| - 2^t(\|x_m\| - \|x_n\|)\} - \|x_m\|$$

$$= \|x_m\| - 2^{t+1}(\|x_m\| - \|x_n\|).$$

57

Thus (5) is fully established for $1 \le r \le k$.

Now observe that $\phi_0, \phi_1, \ldots, \phi_k$ all belong to $U^\perp + V^\perp$. Since $U+V$ is closed, $U^\perp + V^\perp$ is also closed, by 11.3. Hence, by 11.3, there exists a constant c such that each functional ϕ in $U^\perp + V^\perp$ has a representation of the form $\phi = \theta + \psi$, with $\theta \in U^\perp$, $\psi \in V^\perp$, and $\|\theta\| + \|\psi\| \le c\|\phi\|$. We apply this to ϕ_k, getting $\phi_k = \theta + \psi$ and $\|\theta\| + \|\psi\| \le c$. Define $\phi_{k+1} = -\psi - \phi_0$. Obviously $\|\phi_{k+1}\| \le c+1$ and $\phi_{k+1} + \phi_0 = -\psi \in V^\perp$. Also

$$\phi_{k+1} + \phi_k = -\psi - \phi_0 + \theta + \psi = \theta - \phi_0 \in U^\perp.$$

Define $\phi = \sum_{i=0}^{k+1} \phi_i$. Then $\phi \in U^\perp \cap V^\perp$ because

$$\phi = (\phi_0 + \phi_1) + (\phi_2 + \phi_3) + \cdots + (\phi_k + \phi_{k+1}) \in U^\perp$$

and

$$\phi = (\phi_1 + \phi_2) + \cdots + (\phi_{k-1} + \phi_k) + (\phi_{k+1} + \phi_0) \in V^\perp.$$

Since $\|\phi\| \le k + 1 + c + 1 \equiv c'$, we have

$$\text{dist}\,(x, U+V) \ge \phi(x)/c' = \phi(x_m)/c'$$

$$= (c')^{-1} \left[\sum_{i=0}^{k} \phi_i(x_m) + \phi_{k+1}(x_m) \right]$$

$$\ge \frac{k+1}{c'} \{ \|x_m\| - 2^k (\|x_m\| - \|x_{m+k}\|) \} - \frac{c+1}{c'} \|x_m\|$$

Now let $\rho = \lim_{n \to \infty} \|x_n\|$. In the previous inequality, let k be fixed while $m \to \infty$. The result is

$$\text{dist}\,(x, U+V) \ge \frac{k-c}{k+c+2} \rho.$$

Now let $k \to \infty$ to obtain $\text{dist}\,(x, U+V) \ge \rho$. ∎

4.4 THEOREM. *If X is uniformly convex, if $U + V$ is closed, and if A and B are central proximity maps, then the alternating algorithm produces a sequence $\{x_n\}$ such that $\lim_{n \to \infty} (x - x_n)$ is the best approximation of x in $U + V$.*

PROOF. By 4.3, $x - x_n \in U + V$ and $\|x_n\| \downarrow \text{dist}\,(x, U+V)$. Hence $\{x - x_n\}$ is a minimizing sequence in $U + V$ for x. By the uniform convexity, $\lim_{n \to \infty} (x - x_n)$ exists and is the best approximation of x. ∎

4.5 LEMMA. *Let $A : X \longrightarrow U$ be a proximity map such that*

(i) $\|x - 2Ax\| \le \|x\|$ *$(x \in X)$*

(ii) $A(x + u) = Ax + u$ *$(x \in X, u \in U)$.*

Then A is a central proximity map.

58

PROOF. If $u \in U$ and $x \in X$ then by (i) and (ii),

$$\|x - Ax + u\| \geq \|(x - Ax + u) - 2A(x - Ax + u)\|$$
$$= \|x - Ax + u - 2Ax + 2Ax - 2u\|$$
$$= \|x - Ax - u\|.$$

Now repeat this argument with $-u$ in place of u to conclude that

$$\|x - Ax + u\| = \|x - Ax - u\|. \quad \blacksquare$$

4.6 THEOREM. *Every orthogonal projection on Hilbert space is a central proximity map.*

PROOF. Since $x - Ax \perp Ax$, we have by the Pythagoras Theorem

$$\|x\|^2 = \|x - Ax + Ax\|^2 = \|x - Ax\|^2 + \|Ax\|^2 = \|x - Ax - Ax\|^2 = \|x - 2Ax\|^2.$$

Now use 4.5. $\quad \blacksquare$

Notice that 4.6 and 4.4 together give another proof of von Neumann's Theorem, 3.8.

4.7 THEOREM. *A subspace in a Banach space can have at most one central proximity map.*

PROOF. Let A_1 and A_2 be central proximity maps of a Banach space X onto a subspace U. Fix $x \in X$ and put $u = A_1 x - A_2 x$. Consider the equality

$$\|x - A_1 x\| = \|x - A_1 x + ku\|.$$

This equality is clearly true for $k = 0$. If it is true for $k = n$, then

$$\|x - A_1 x\| = \|x - A_1 x + nu\| = \|x - A_2 x + (n+1)u\|$$
$$= \|x - A_2 x - (n+1)u\| = \|x - A_1 x - (n+2)u\| = \|x - A_1 x + (n+2)u\|.$$

Thus, by induction we have

$$\|x - A_1 x\| = \|x - A_1 x + 2nu\| \geq 2n \|u\| - \|x - A_1 x\|.$$

Letting $n \rightarrow \infty$, we see that $\|u\| = 0$. $\quad \blacksquare$

CHAPTER 5

THE DILIBERTO-STRAUS ALGORITHM IN $C(S \times T)$

In this chapter, a particular instance of the alternating algorithm will be considered in detail. The setting is the space $C(S \times T)$ of continuous functions on $S \times T$, where S and T are compact Hausdorff spaces. The two subspaces which figure in the alternating algorithm are $C(S)$ and $C(T)$. Here we identify an element $u \in C(S)$ with an element $\bar{u} \in C(S \times T)$ by writing $\bar{u}(s, t) = u(s)$. Henceforth we do not belabor this distinction.

The first investigation of the alternating algorithm in this case was carried out by Diliberto and Straus in [56]. Their work was independent of von Neumann's, and the results and methods are quite different; only the algorithm itself is the same. Thus it seems appropriate to refer to the algorithm by the names of Diliberto and Straus. The question of convergence of the algorithm was left open until the work of Aumann [6].

The simplest and most natural proximity maps of $C(S \times T)$ onto $C(S)$ and $C(T)$ are given by

$$(Az)(s) = \tfrac{1}{2} \max_t z(s, t) + \tfrac{1}{2} \min_t z(s, t)$$
$$(Bz)(t) = \tfrac{1}{2} \max_s z(s, t) + \tfrac{1}{2} \min_s z(s, t).$$

With these maps in hand, one can then define a sequence $\{z_n\}$, starting with any $z_0 \in C(S \times T)$, by the formulae

$$z_{2n} = z_{2n-1} - Az_{2n-1}$$
$$z_{2n+1} = z_{2n} - Bz_{2n}.$$

Diliberto and Straus proved, among other things, that $\|z_n\| \downarrow \mathrm{dist}\,(z, C(S) + C(T))$ and that the sequence $\{z_n\}$ has cluster points. Aumann subsequently proved that the sequence $\{z - z_n\}$ converges to a best approximation of z in $C(S) + C(T)$.

It will be convenient to define an "averaging functional" $\mathcal{M}$ on univariate functions by writing

$$\mathcal{M}f = \tfrac{1}{2} \sup_s f(s) + \tfrac{1}{2} \inf_s f(s).$$

We can then express A and B as follows:

$$(Az)(s) = \mathcal{M}z_s \quad \text{and} \quad (Bz)(t) = \mathcal{M}z^t$$

where z_s and z^t are the "sections" defined by $z_s(t) = z^t(s) = z(s, t)$.

5.1 LEMMA. *The maps A and B just defined are non-expansive, central, proximity maps.*

PROOF. For a function $x \in C(S)$, the constant which best approximates x is given by $\alpha = \mathcal{M}x$, or

$$\alpha = \tfrac{1}{2} \max_s x(s) + \tfrac{1}{2} \min_s x(s).$$

Therefore, the operator B has the property that for any $y \in C(T)$ and for any fixed t,

$$\sup_s |z(s,t) - (Bz)(t)| \leq \sup_s |z(s,t) - y(t)|.$$

Consequently

$$\|z - Bz\| \leq \|z - y\|.$$

This shows that B is a proximity map of $C(S \times T)$ onto $C(T)$.

It is apparent that B is order-preserving and that $B(z + r) = Bz + r$ for any constant r. Hence if $r = \|z - w\|$ then from the pointwise inequality

$$-r + w \leq z \leq r + w$$

we conclude that

$$-r + Bw \leq Bz \leq r + Bw$$

whence

$$\|Bz - Bw\| \leq r = \|z - w\|.$$

This shows that B is non-expansive. In order to prove that B has the centrality property, it suffices to show that for arbitrary $z \in C(S \times T)$ and $u \in C(T)$,

$$\|z - Bz - u\| \geq \|z - Bz + u\|.$$

Let $w = z - Bz$, and select (s_0, t_0) so that

$$\|w + u\| = \sigma[(w + u)(s_0, t_o)] \qquad \sigma = \pm 1.$$

Since $Bw = 0$, there is a point s_1 such that

$$\sigma[w(s_0, t_0) + w(s_1, t_0)] \leq 0.$$

Indeed, let s_1 be a maximum point of $w(s, t_0)$ if $\sigma < 0$, and let s_1 be a minimum point if $\sigma > 0$. Now we have

$$\|w - u\| \geq \sigma[u(t_0) - w(s_1, t_0)] \geq \sigma[u(t_0) + w(s_0, t_0)] = \|w + u\|. \quad \blacksquare$$

5.2 LEMMA. *The averaging functional* M *is non-expansive:*

$$|Mf_1 - Mf_2| \le \|f_1 - f_2\|.$$

PROOF. If $\alpha = \|f_1 - f_2\|$ then

$$f_2 - \alpha \le f_1 \le f_2 + \alpha.$$

By using obvious properties of M, we conclude that

$$Mf_2 - \alpha \le Mf_1 \le Mf_2 + \alpha.$$

This implies the inequality to be proved. ∎

The alternating algorithm in the present setting will be termed the "Diliberto-Straus Algorithm." Using the operators A and B described above, we write it in the following form:

$$\begin{cases} z_0 \in C(S \times T) & z_{n+1} = z_n - w_n \\ w_n = Az_n & \text{if } n \text{ is odd,} \\ w_n = Bz_n & \text{if } n \text{ is even.} \end{cases}$$

We will prove below that the sequence $z_0 - z_n$ converges uniformly to a best approximation of z_0 in $C(S) + C(T)$.

5.3 LEMMA. *For two continuous functions,* f *and* g, *on a a compact domain,*

$$\sup f - \sup g \le \sup(f - g).$$

If equality occurs here, then there exists a single point where all three suprema are attained simultaneously.

PROOF. For each s we have

(1) $f(s) = f(s) - g(s) + g(s) \le \sup(f - g) + g(s) \le \sup(f - g) + \sup g$

(2) $f(s) = f(s) - g(s) + g(s) \le f(s) - g(s) + \sup g \le \sup(f - g) + \sup g.$

Either of these yields the inequality

$$\sup f \le \sup(f - g) + \sup g.$$

Suppose now that equality holds in this inequality. Let ξ be a point such that $f(\xi) = \sup f$. From (1),

$$\sup(f - g) + \sup g = \sup f = f(\xi) \le \sup(f - g) + g(\xi).$$

Hence $g(\xi) = \sup g$. From (2),

$$\sup(f - g) + \sup g = \sup f = f(\xi) \le f(\xi) - g(\xi) + \sup g.$$

Hence $f(\xi) - g(\xi) = \sup(f - g)$. ∎

We come now to one of the crucial lemmas established by Aumann. Having fixed z_0, we define the sequence

$$\lambda_n = \max\{|z_n(s, t)| : |w_n(s, t)| = \|w_n\|\}.$$

5.4 LEMMA. If $\|w_n\| = \|w_{n-1}\|$ then $\lambda_{n-1} \geq \lambda_n + \|w_n\|$.

PROOF. Suppose that n is even; the other case is similar. Since B is non-expansive and $Bz_{n-1} = 0$, we have

$$\|w_n\| = \|Bz_n\| = \|Bz_n - Bz_{n-1}\| \leq \|z_n - z_{n-1}\| = \|w_{n-1}\| = \|w_n\|.$$

By the definition of λ_n, there exists a point $\tau \in T$ such that

$$|w_n(\tau)| = \|w_n\| \quad \text{and} \quad \max_s |z_n(s, \tau)| = \lambda_n.$$

There are now two cases, depending on the sign of $w_n(\tau)$. We assume that $w_n(\tau) = \|w_n\|$. Since $w_n = Bz_n - Bz_{n-1}$, we have

$$\tfrac{1}{2}[\max_s z_n(s, \tau) + \min_s z_n(s, \tau)] - \tfrac{1}{2}[\max_s z_{n-1}(s, \tau) + \min_s z_{n-1}(s, \tau)] = \|w_{n-1}\|.$$

This can be rewritten as follows:

$$\tfrac{1}{2}[\max_s z_n(s, \tau) - \max_s z_{n-1}(s, \tau)] + \tfrac{1}{2}[\max_s -z_{n-1}(s, \tau) - \max_s -z_n(s, \tau)] = \|w_{n-1}\|.$$

The bracketed expressions in the preceding equation do not exceed $\|w_{n-1}\|$, as we see with the help of 5.3:

$$\max_s z_n(s, \tau) - \max_s z_{n-1}(s, \tau) \leq \max_s[z_n(s, \tau) - z_{n-1}(s, \tau)]$$
$$= \max_s[-w_{n-1}(s, \tau)] \leq \|w_{n-1}\|.$$

The other term is analyzed similarly. We can now conclude that

$$\max_s z_n(s, \tau) - \max_s z_{n-1}(s, \tau) = \|w_{n-1}\|$$
$$\max_s[z_n(s, \tau) - z_{n-1}(s, \tau)] = \|w_{n-1}\|$$
$$\max_s[-w_{n-1}(s, \tau)] = \|w_{n-1}\|.$$

By the second half of 5.3, we conclude that there exists a point σ such that

$$-w_{n-1}(\sigma) = \|w_{n-1}\|$$
$$z_n(\sigma, \tau) = \max_s z_n(s, \tau)$$
$$z_{n-1}(\sigma, \tau) = \max_s z_{n-1}(s, \tau).$$

By previous equations,

$$0 \leq \|w_n\| = w_n(\tau) = (Bz_n)(\tau) = \tfrac{1}{2} \max_s z_n(s, \tau) + \tfrac{1}{2} \min_s z_n(s, \tau).$$

63

Consequently,

$$\max_s z_n(s, \tau) \geq - \min_s z_n(s, \tau)$$

and

$$\lambda_n = \max_s |z_n(s, \tau)| = \max_s z_n(s, \tau) = z_n(\sigma, \tau).$$

Using the equation

$$w_{n-1}(\sigma) = (Az_{n-1})(\sigma) = \tfrac{1}{2} \max_t z_{n-1}(\sigma, t) + \tfrac{1}{2} \min_t z_{n-1}(\sigma, t)$$

we now infer that

$$\begin{aligned}
\lambda_{n-1} &\geq - \min_t z_{n-1}(\sigma, t) \\
&= \max_t z_{n-1}(\sigma, t) - 2w_{n-1}(\sigma) \\
&\geq z_{n-1}(\sigma, \tau) - w_{n-1}(\sigma) - w_{n-1}(\sigma) \\
&= z_n(\sigma, \tau) + \|w_{n-1}\| \\
&= \lambda_n + \|w_{n-1}\|.
\end{aligned}$$

This completes the proof in the case $w_n(\tau) = \|w_n\|$. In the other case, we write

$$-w_n(\tau) = (Bz_{n-1} - Bz_n)(\tau)$$

and proceed as before. We find a point σ such that

$$\begin{aligned}
w_{n-1}(\sigma) &= \|w_{n-1}\|, \\
z_{n-1}(\sigma, \tau) &= \min_s z_{n-1}(s, \tau),
\end{aligned}$$

and

$$z_n(\sigma, \tau) = \min_s z_n(s, \tau).$$

Then

$$\begin{aligned}
\lambda_{n-1} &\geq \max_t z_{n-1}(\sigma, t) \geq -z_n(\sigma, \tau) + \|w_{n-1}\| \\
&\geq \lambda_n + \|w_{n-1}\|. \quad \blacksquare
\end{aligned}$$

For the remaining parts of the analysis, we introduce some convenient notation:

$$v_n = w_0 + w_2 + w_4 + \cdots + w_{2n}$$
$$u_n = w_1 + w_3 + w_5 + \cdots + w_{2n-1}.$$

Then it follows that $v_n \in C(T)$, $u_n \in C(S)$, $z_{2n} = z - u_n - v_{n-1}$, and $z_{2n+1} = z - u_n - v_n$.

Define operators $A' : V \to U$ and $B' : U \to V$ by putting

$$A'v = A(z - v), \qquad B'u = B(z - u)$$

where z is the fixed member of $C(S \times T)$ which is the starting point of the Diliberto-Straus iteration. The properties of A give us

$$0 = A(z_{2n-1} - Az_{2n-1}) = Az_{2n} = A(z - u_n - v_{n-1}) = A(z - v_{n-1}) - u_n.$$

Hence $u_n = A(z - v_{n-1}) = A'v_{n-1}$. Similarly, $v_n = B'u_n$. Hence $u_{n+1} = A'v_n = A'B'u_n$ and $v_{n+1} = B'u_{n+1} = B'A'v_n$.

5.5 LEMMA. *For all n,*

$$|u_n(s) - u_n(\sigma)| \le \|z_s - z_\sigma\|$$
$$|v_n(t) - v_n(\tau)| \le \|z^t - z^\tau\|$$
$$|z_n(s,t) - z_n(\sigma,\tau)| \le 2\|z_s - z_\sigma\| + 2\|z^t - z^\tau\|.$$

PROOF. As noted above, $u_n = A(z - v_{n-1})$. With the help of the averaging functional M introduced previously, we have

$$|u_n(s) - u_n(\sigma)| = |[A(z - v_{n-1})](s) - [A(z - v_{n-1})](\sigma)|$$
$$= |M(z_s - v_{n-1}) - M(z_\sigma - v_{n-1})| \le \|(z_s - v_{n-1}) - (z_\sigma - v_{n-1})\|$$
$$= \|z_s - z_\sigma\|.$$

The proof for v_n is similar. As for z_{2n}, we have $z_{2n} = z - u_n - v_{n-1}$, whence

$$|z_{2n}(s,\tau) - z_{2n}(\sigma,\tau)| \le |z_{2n}(s,t) - z_{2n}(\sigma,t)| + |z_{2n}(\sigma,t) - z_{2n}(\sigma,\tau)|$$
$$= |z(s,t) - u_n(s) - z(\sigma,t) + u_n(\sigma)| + |z(\sigma,t) - v_{n-1}(t) - z(\sigma,\tau) + v_{n-1}(\tau)|$$
$$\le |z(s,t) - z(\sigma,t)| + |u_n(s) - u_n(\sigma)| + |z(\sigma,t) - z(\sigma,\tau)| + |v_{n-1}(t) - v_{n-1}(\tau)|$$
$$\le 2\|z_s - z_\sigma\| + 2\|z^t - z^\tau\|.$$

The proof for z_{2n+1} is similar. ∎

The next result is the second of the crucial lemmas due to Aumann.

5.6 LEMMA. *In the Diliberto-Straus algorithm, we have $\lim w_n = 0$.*

PROOF. We have seen that $\|w_{n+1}\| \le \|w_n\|$ for all n. (See the first equation in the proof of 5.4.) Hence we may define $\epsilon = \lim \|w_n\|$. Because of equicontinuity of $\{z_n\}$ (5.5) there is a uniformly convergent subsequence, $z_{n_k} \to z^*$. If the algorithm is applied to z^* as starting point, the result is a sequence which we denote by $z_1^*, z_2^*, \ldots$. Since the operations in the algorithm are continuous, we have $z_m^* = \lim_k(z_{n_k+m})$. It follows that $\|z_{m+1}^* - z_m^*\| = \lim \|z_{n_k+m+1} - z_{n_k+m}\| = \lim \|w_{n_k+m}\| = \epsilon$. By 5.4, applied to z_m^*, we have $\lambda_m^* \le \lambda_{m-1}^* - \epsilon$ for $m = 1, 2, \ldots$. This leads to $0 \le \lambda_m^* \le \lambda_1^* - (m-1)\epsilon$, and therefore to the conclusion that $\epsilon = 0$. ∎

5.7 LEMMA. *The sequence $\{u_n\}$ is bounded.*

PROOF. By 5.5 the sequence $\{z_n\}$ is equicontinuous. It is certainly bounded, since $\|z_0\| \geq \|z_1\| \geq \cdots$. By the Ascoli Theorem, the sequence $\{z_{2n}\}$ has a convergent subsequence, $\{z_{2n_k}\}$. Since W is closed and $z_n - z \in W$, we can write $\lim z_{2n_k} = z - u - v$ for appropriate $u \in C(S)$ and $v \in C(T)$. Since $Bz_{2n} = 0$, we have

$$0 = B(z - u - v) = B(z - u) - v = B'u - v.$$

Thus $v = B'u$. Since $z_{n+1} - z_n \to 0$ by 5.6, we have $z_{2n_k+1} \to z - u - v$. From the equation $Az_{2n+1} = 0$ we conclude as above that $u = A'v$. Hence $u = A'B'u$. The boundedness of $\{u_n\}$ now follows from the non-expansiveness of $A'B'$:

$$\|u_{n+1} - u\| = \|A'B'u_n - A'B'u\| \leq \|u_n - u\|. \quad \blacksquare$$

5.8 THEOREM. *The sequence $\{z_n\}$ produced from an arbitrary $z \in C(S \times T)$ by the Diliberto-Straus Algorithm converges uniformly, and $z - \lim z_n$ is a best approximation of z in $C(S) + C(T)$.*

PROOF. By 5.5, the sequence $\{u_n\}$ is equicontinuous. By 5.7 it is bounded. Hence by the Ascoli Theorem there exists a convergent subsequence, $\{u_{n_k}\}$. Put $u^* = \lim_k u_{n_k}$. Then (by an equation just prior to 5.5) $u_{n_k+1} = A'B'u_{n_k} \to A'B'u^*$. By 5.6,

$$\|u_{n+1} - u_n\| = \|w_{2n+1}\| \to 0.$$

Therefore $u_{n_k+1} \to u^*$ and $A'B'u^* = u^*$. As in 5.7, we have

$$\|u_{n+1} - u^*\| = \|A'B'u_n - A'B'u^*\| \leq \|u_n - u^*\|.$$

This shows that $u_n \to u^*$. Hence $v_n = B'u_n \to B'u^* \equiv v^*$. By 4.3, $\|z_{2n}\| \downarrow \text{dist}\,(z, W)$. Hence

$$\|z - u_n - v_{n-1}\| \downarrow \text{dist}\,(z, W)$$

and

$$\|z - u^* - v^*\| = \text{dist}\,(z, W). \quad \blacksquare$$

CHAPTER 6

THE ALGORITHM OF VON GOLITSCHEK

In a recent series of papers [69-74], von Golitschek has developed a powerful new algorithm for obtaining sup-norm approximations of the form

$$(1) \qquad z(s,t) \approx f[x(s)h(t) + y(t)g(s)]$$

in which g, h, f, z are all prescribed continuous functions, and the functions x and y are sought. His algorithm provides a constructive proof of the following theorem:

6.1 THEOREM. *Let S and T be compact Hausdorff spaces. Let*

$$z \in C(S \times T), \ \ g \in C(S), \ \ h \in C(T), \ \ g > 0, \quad \text{and} \quad h > 0.$$

Let f be a strictly increasing element of $C(\mathbb{R})$ such that $f^{-1} \in C(\mathbb{R})$. Then z has a best approximation of the form (1), with $x \in C(S)$ and $y \in C(T)$.

In fact, von Golitschek proves 6.1 for $z \in C(D)$, where D is a subset of $S \times T$ subject to some technical hypotheses. Also he proves 6.1 with somewhat less restrictive conditions on f. We prove Theorem 6.1 below.

In what follows, z, g, h, f remain fixed, and we set

$$W = \{f \circ (xh + yg) \ : \ x \in C(S), y \in C(T)\}.$$

Observe that W is not a linear subspace. However, since f is invertible, some techniques of linear approximation theory are applicable.

The von Golitschek Algorithm contains a real parameter α chosen to lie in the interval $0 \le \alpha \le \|z - f \circ 0\|$. If $\alpha > \text{dist}(z, W)$, then in a finite number of steps the algorithm will produce an element $w \in W$ satisfying $\|z - w\| \le \alpha$.

Having fixed α, we define

$$k(s,t) = f^{-1}[z(s,t) - \alpha]/g(s)h(t)$$

$$K(s,t) = f^{-1}[z(s,t) + \alpha]/g(s)h(t).$$

The algorithm starts by defining

$$x_0(s) = 0 \quad \text{and} \quad y_0(t) = \inf_s K(s, t).$$

At the ith step, we define

$$x_i(s) = x_{i-1}(s) \vee \sup_t [k(s, t) - y_{i-1}(t)]$$

and

$$y_i(t) = y_{i-1}(t) \wedge \inf_s [K(s, t) - x_i(s)].$$

If $y_i = y_{i-1}$, then STOP. (The symbols $\vee$ and $\wedge$ denote the pointwise maximum and minimum operations, respectively.)

6.2 LEMMA. *If t_i is a point such that $y_i(t_i) \leq y_{i-1}(t_i)$, then there exist points s_i and t_{i-1} such that*

(i) $y_i(t_i) = y_{i-1}(t_{i-1}) + K(s_i, t_i) - k(s_i, t_{i-1})$

(ii) $y_{i-1}(t_{i-1}) < y_{i-2}(t_{i-1})$.

PROOF. Let s_i be a minimum point for the function $K(\bullet, t_i) - x_i$. Let t_{i-1} be a maximum point for the function $k(s_i, \bullet) - y_{i-1}$. Then

$$K(s_i, t_i) - x_i(s_i) = \inf_s [K(s, t_i) - x_i(s)]$$
$$= y_i(t_i) < y_{i-1}(t_i)$$
$$\leq K(s_i, t_i) - x_{i-1}(s_i).$$

Hence, $x_{i-1}(s_i) < x_i(s_i)$. Similarly, we have

$$k(s_i, t_{i-1}) - y_{i-1}(t_{i-1}) = \sup_t [k(s_i, t) - y_{i-1}(t)]$$
$$= x_i(s_i) > x_{i-1}(s_i)$$
$$\geq k(s_i, t_{i-1}) - y_{i-1}(t_{i-1}).$$

Hence, $y_{i-1}(t_{i-1}) < y_{i-2}(t_{i-1})$. This verifies part (ii). The verification of (i) is now as follows:

$$y_i(t_i) = K(s_i, t_i) - x_i(s_i)$$
$$= K(s_i, t_i) - k(s_i, t_{i-1}) + y_{i-1}(t_{i-1}). \quad \blacksquare$$

6.3 LEMMA. *If $n \geq 1$ and if t_n is a point such that $y_n(t_n) < y_{n-1}(t_n)$ then there exists a "path"*

$$(s_1, t_0), (s_1, t_1), (s_2, t_1), (s_2, t_2), \cdots, (s_n, t_n)$$

such that

$$y_n(t_n) = \sum_{i=1}^n \{K(s_i, t_i) - k(s_i, t_{i-1})\} + y_0(t_0).$$

PROOF. Use the preceding lemma n times. $\blacksquare$

68

6.4 DEFINITION. *A path is an ordered set of points*

$$(s_1, t_0), (s_1, t_1), (s_2, t_1), (s_2, t_2), \cdots, (s_n, t_{n-1}), (s_n, t_n).$$

The path is said to be closed if $t_n = t_0$.

6.5 THEOREM. *If the algorithm stops in the nth step, then $\alpha \geq \text{dist}(z, W)$ and $\|z - f \circ (ghx_n + ghy_n)\| \leq \alpha$. If $\alpha > \text{dist}(z, W)$ then the algorithm will stop at some step.*

PROOF. The definition of y_i yields the following pointwise inequality:

$$y_i \leq K - x_i = \frac{f^{-1} \circ (z + \alpha)}{gh} - x_i.$$

When this is rearranged, it reads

$$-\alpha \leq z - f \circ (x_i gh + y_i gh).$$

Similarly, starting with the definition of x_i, we obtain

$$z - f \circ (x_i gh + y_{i-1} gh) \leq \alpha.$$

It follows that in the case $y_n = y_{n-1}$ we will have

$$-\alpha \leq z - f \circ (x_n gh + y_n gh) \leq \alpha.$$

This shows that $\|z - f \circ (x_n gh + y_n gh)\| \leq \alpha$ and that $\alpha \geq \rho \equiv \text{dist}(z, W)$. Now we want to prove that if $\alpha \geq \rho$, then for some n, $y_n = y_{n-1}$. If this conclusion is false, then for an arbitrary integer n there exists a point $t_n \in T$ such that $y_n(t_n) < y_{n-1}(t_n)$. Then by 6.3, there exists a path

$$(s_1, t_0), (s_1, t_1), (s_2, t_1), (s_2, t_2), \cdots, (s_n, t_n)$$

such that

$$y_n(t_n) - y_0(t_0) = \sum_{i=1}^{n} \{K(s_i, t_i) - k(s_i, t_{i-1})\}.$$

Select β so that $\rho < \beta < \alpha$. By compactness and continuity,

$$\frac{f^{-1} \circ (z + \alpha)}{gh} - \frac{f^{-1} \circ (z + \beta)}{gh} \geq \epsilon$$

and

$$\frac{f^{-1} \circ (z - \beta)}{gh} - \frac{f^{-1} \circ (z - \alpha)}{gh} \geq \epsilon$$

for some $\epsilon > 0$. Since $\beta > \rho$, there exist $x \in C(S)$ and $y \in C(T)$ such that

$$\|z - f \circ (xgh + ygh)\| < \beta.$$

This leads to the inequality

$$-\beta < f \circ (xgh + ygh) - z < \beta$$

and then to

$$\frac{f^{-1} \circ (z - \beta)}{gh} < x + y < \frac{f^{-1} \circ (z + \beta)}{gh}.$$

It follows that

$$k + \epsilon = \frac{f^{-1} \circ (z - \alpha)}{gh} + \epsilon < x + y < \frac{f^{-1} \circ (z + \alpha)}{gh} - \epsilon = K - \epsilon.$$

From a previous equation, therefore,

$$y_n(t_n) - y_0(t_0) \geq \sum_{i=1}^{n} \{[x(s_i) + y(t_{i-1}) + \epsilon] - [x(s_i) + y(t_{i-1}) - \epsilon]\}$$
$$= y(t_n) - y(t_0) + 2n\epsilon \;\; \to \infty \;\; \text{as } n \to \infty.$$

This however is a contradiction, since $y_0 \geq y_1 \geq \cdots$ directly from the definition of the algorithm. ∎

6.6 LEMMA. Let $F \in C(S \times T)$. Define $F^-(s) = \inf_t F(s, t)$ and $F^+(s) = \sup_t F(s, t)$. Then for $s, \sigma \in S$ we have

$$|F^-(s) - F^-(\sigma)| \leq \|F_s - F_\sigma\| \quad \text{and} \quad |F^+(s) - F^+(\sigma)| \leq \|F_s - F_\sigma\|$$

where F_s is the s-section of F, defined by the equation $F_s(t) = F(s, t)$.

PROOF. For every $s, \sigma,$ and t,

$$F^-(s) \leq F(s, t) = F(s, t) - F(\sigma, t) + F(\sigma, t) \leq \|F_s - F_\sigma\| + F(\sigma, t).$$

By taking an infimum we get

$$F^-(s) \leq \|F_s - F_\sigma\| + F^-(\sigma).$$

This establishes that

$$F^-(s) - F^-(\sigma) \leq \|F_s - F_\sigma\|.$$

By interchanging s and σ, we get

$$F^-(\sigma) - F^-(s) \leq \|F_\sigma - F_s\|.$$

The last two inequalities imply the one in the lemma. The inequality for F^+ is proved in the same way. ∎

6.7 LEMMA. *For each n, s, σ, t, τ we have*

$$|x_n(s) - x_n(\sigma)| \le \|k_s - k_\sigma\|$$

and

$$|y_n(t) - y_n(\tau)| \le \|K^t - K^\tau\|.$$

PROOF. Since $x_0(s) = 0$, the first inequality is trivial when $n = 0$. Since

$$y_0(t) = \inf_s K(s, t)$$

the second inequality follows from 6.6 when $n = 0$. If the two desired inequalities are assumed true for an index n, then for the index $n + 1$ we have

$$x_{n+1}(s) = x_n(s) \vee \sup_t [k(s, t) - y_n(t)].$$

By 6.6 (specialized to a two-point set!) we see that we need only verify these two inequalities:

(2) $$|x_n(s) - x_n(\sigma)| \le \|k_s - k_\sigma\|$$

(3) $$\left| \sup_t [k(s, t) - y_n(t)] - \sup_t [k(\sigma, t) - y_n(t)] \right| \le \|k_s - k_\sigma\|.$$

The first of these is our induction hypothesis. The second is established by using 6.6. Indeed, the left side of (3) does not exceed

$$\|(k - y_n)_s - (k - y_n)_\sigma\| = \|k_s - k_\sigma\|.$$

The second inequality in the lemma, for the case $n + 1$, is proved in the same way. ∎

6.8 LEMMA. *If $\alpha \ge \mathrm{dist}(z, W)$, then for each closed path,*

$$\sum_{i=0}^{n} [K(s_i, t_i) - k(s_i, t_{i-1})] \ge 0.$$

PROOF. We prove the lemma under the assumption that $\alpha > \rho = \mathrm{dist}(z, W)$. The case $\alpha = \rho$ will then follow by continuity of K and k as functions of α. If $\alpha > \rho$, then there

71

exist $x \in C(S)$ and $y \in C(T)$ such that $\|z - f \circ (xgh + ygh)\| \leq \alpha$. This is equivalent to each of the following inequalities:

$$-\alpha \leq f \circ (xgh + ygh) - z \leq \alpha$$

$$z - \alpha \leq f \circ (xgh + ygh) \leq z + \alpha$$

$$f^{-1} \circ (z - \alpha) \leq xgh + ygh \leq f^{-1} \circ (z + \alpha)$$

$$[f^{-1} \circ (z - \alpha)]/gh \leq x + y \leq [f^{-1} \circ (z + \alpha)]/gh$$

$$k \leq x + y \leq K.$$

Now define two linear functionals on $C(S \times T)$ which are sums of point functionals

$$\phi = \sum_{i=0}^{n} (s_i, t_i)^{\wedge} \quad \text{and} \quad \psi = \sum_{i=0}^{n} (s_i, t_{i-1})^{\wedge}.$$

It is easily seen that $\phi - \psi \perp C(S) + C(T)$. Also, ϕ and ψ are nonnegative functionals. Hence,

$$\phi(x + y) \leq \phi(K) \quad \text{and} \quad \psi(k) \leq \psi(x + y).$$

This is the inequality which was to be proved. ∎

6.9 LEMMA. *If the parameter* α *in the algorithm satisfies* $\alpha \geq \mathrm{dist}(z, W)$, *then* $y_n \geq C \equiv -2\|K\| - \|k\|$ *for all* n.

PROOF. It is clear that $y_0 \geq C$. If the set

$$J_n = \{t \; : \; y_n(t) < y_{n-1}(t)\}$$

is nonempty, and if $\tau \in J_n$ then by 6.3, there is a path for which $\tau = t_n$ and

$$y_n(\tau) = y_0(t_0) + \sum_{i=1}^{n} [K(s_i, t_i) - k(s_i, t_{i-1})].$$

By adding two appropriate points, (s_0, t_0) and (s_0, t_{-1}), we create a closed path. Then by 6.8,

$$y_n(\tau) = y_0(t_0) + \sum_{i=0}^{n} [K(s_i, t_i) - k(s_i, t_{i-1})] - K(s_0, t_0) + k(s_0, t_{-1})$$

$$\geq y_0(t_0) - K(s_0, t_0) + k(s_0, t_{-1})$$

$$\geq -\|y_0\| - \|K\| - \|k\| \geq -2\|K\| - \|k\| \equiv C.$$

From this it can be proved that $y_n \geq C$. In order to do so, suppose on the contrary that $y_n(\xi) < C$ for some ξ. By the previous analysis, it follows that $\xi \notin J_n$. Hence

$y_{n-1}(\xi) = y_n(\xi) < C$. By repeating this argument n times we arrive at the conclusion that $y_0(\xi) < C$, which is absurd.

If one of the sets J_n is empty, then the algorithm stops. The preceding J_{n-1} is nonempty, and the previous argument shows that $y_{n-1} \geq C$. Since $y_0 \geq y_1 \geq \ldots \geq y_{n-1} = y_n$ the proof is complete. ∎

6.10 THEOREM. *If the parameter α in the algorithm is set equal to* $\mathrm{dist}(z, W)$, *then either the algorithm terminates and yields a solution as in 6.5, or it produces sequences* $\{x_n\}$ *and* $\{y_n\}$ *which converge uniformly and monotonically to functions* $x \in C(S)$ *and* $y \in C(T)$ *for which* $\|z - f \circ (ghx + ghy)\| = \mathrm{dist}(z, W)$.

PROOF. Theorem 6.5 takes care of the terminating case. In the other case we have

$$x_0 \leq x_1 \leq x_2 \leq \ldots \quad \text{and} \quad y_0 \geq y_1 \geq y_2 \geq \ldots.$$

By 6.9, $y_n \geq C$ for all n.

The sequence $\{y_n\}$, being bounded from below and nonincreasing, converges pointwise to a function $y \in \ell_\infty(T)$. Since the sequence is equicontinuous (by 6.7), we have $y \in C(T)$.

In order to see that the sequence $\{x_n\}$ is bounded from above, we recall an inequality from the proof of 6.5:

$$-\alpha \leq z - f \circ (x_n gh + y_n gh).$$

From this we obtain

$$x_n + y_n \leq [f^{-1} \circ (z + \alpha)]/gh = K.$$

Since $\{y_n\}$ is bounded from below, $\{x_n\}$ is bounded from above. We now conclude that the sequence $\{x_n\}$ converges uniformly and monotonically (upward) to an element x of $C(S)$.

In order to see that xg and yh provide a solution to the approximation problem, take the limit in an inequality above to get

$$-\alpha \leq z - f \circ (xgh + ygh).$$

Likewise, from another inequality in the proof of 6.5, we have in the limit

$$z - f \circ (xgh + ygh) \leq \alpha.$$

Since $\alpha = \mathrm{dist}(z, W)$, this yields $\|z - f \circ (xgh + ygh)\| \leq \mathrm{dist}(z, W)$. ∎

The preceding theorem establishes the existence of a solution to the approximation problem posed at the beginning, viz., to find $x^* \in C(S)$ and $y^* \in C(T)$ to minimize the deviation

$$\sup_{s,t}\left|z(s,t) - f\big(x^*(s)h(t) + y^*(t)g(s)\big)\right|.$$

The solution pair is defined by $x^* = xg$, $y^* = yh$, where x and y are the limit functions from the von Golitschek Algorithm.

It should be noted that the constructive proof given by the algorithm establishes the existence of a solution pair x^*, y^* with the property that x^*/g has the same modulus of continuity as k, and y^*/h has the same modulus of continuity as K. (See 6.7.)

In practice the minimum deviation ρ is not known in advance. A binary search procedure can be used to determine ρ, or a close estimate of it. For this purpose, the algorithm should be modified by adding another stopping criterion; viz., in step i, stop if $\inf_t y_i(t) < C \equiv -2\|K\| - \|k\|$. If this stopping condition is met, it signifies that $\alpha < \rho$, in accordance with 6.9. On the other hand, if $\alpha < \rho$ then this stopping criterion will be met at some stage, in accordance with the following lemma.

6.11 LEMMA. *If the parameter α in the algorithm satisfies $\alpha < \mathrm{dist}(z, W)$, then at some step, $\inf_t y_n(t) < C$.*

PROOF. If $y_n \geq C$ for all n, then as in the proof of 6.10 the sequences x_n and y_n converge to functions $x \in C(S)$ and $y \in C(T)$ such that

$$\|z - f \circ (ghx + ghy)\| \leq \alpha.$$

Hence $\alpha \geq \rho$. ∎

The binary search algorithm starts with the interval $[0, \|z - f \circ 0\|]$. In the general step, an interval $[a, b]$ will be available from the previous step. It will satisfy $0 \leq a \leq \rho \leq b \leq \|z - f \circ 0\|$. Then we set $\alpha = \frac{1}{2}(a + b)$ and apply the algorithm with this value of α. If $\alpha < \rho$, the algorithm (with both stopping criteria) will stop with $\inf_t y_n(t) < C$. If $\alpha > \rho$, the algorithm will stop with $y_n = y_{n-1}$. If $\alpha = \rho$, the algorithm may not stop, but will produce a solution in the limit. In the first case of stopping, we replace $[a, b]$ by $[\alpha, b]$. In the second case of stopping we replace $[a, b]$ by $[a, \alpha]$. In both of these cases we begin afresh with the new interval.

CHAPTER 7

THE L_1-VERSION OF THE DILIBERTO-STRAUS ALGORITHM

In this chapter, S and T denote compact Hausdorff spaces. As usual, $C(S)$ and $C(T)$ are the spaces of real-valued continuous functions on S and T respectively, with supremum norms written as $\| \ \|_\infty$. We assume, in addition, that regular Borel measures μ and ν have been prescribed on S and T respectively. The product measure on $S \times T$ is denoted by σ. It is assumed that $\mu(S) = \nu(T) = 1$. Our interest now is in L_1-approximation problems in $C(S \times T)$, where the various L_1-norms are

$$\|x\|_1 = \int |x(s)| \, d\mu \qquad x \in C(S)$$
$$\|y\|_1 = \int |y(t)| \, d\nu \qquad y \in C(T)$$
$$\|z\|_1 = \int |z(s,t)| \, d\sigma \qquad z \in C(S \times T).$$

Let G and H be finite-dimensional subspaces in $C(S)$ and $C(T)$ respectively. Then there exist L_1-proximity maps

$$A^0 : C(S) \twoheadrightarrow G, \qquad B^0 : C(T) \twoheadrightarrow H.$$

Thus for all $x \in C(S)$, $g \in G$, $y \in C(T)$, and $h \in H$,

$$\|x - A^0 x\|_1 \leq \|x - g\|_1 \text{ and } \|y - B^0 y\|_1 \leq \|y - h\|_1.$$

These maps are extended in the standard way to $C(S \times T)$. Denoting the extensions by A and B, we have then

$$(Az)(s,t) = (A^0 z^t)(s) \qquad (Bz)(s,t) = (B^0 z_s)(t).$$

Here z_s and z^t are the sections defined by

$$z^t(s) = z_s(t) = z(s,t).$$

Note that in general Az will not belong to $C(S \times T)$ unless we assume the continuity of the map A^0.

7.1 LEMMA. Let $A^0 : C(S) \twoheadrightarrow G$ be an L_1-proximity map onto a finite-dimensional subspace G. If A^0 is continuous in the supremum norm then A (as defined above) is an L_1-proximity map of $C(S \times T)$ onto $G \otimes C(T)$, and is continuous in the supremum norm.

PROOF. That A is a continuous map of $C(S \times T)$ into itself is established by 11.9. The images Az lie in $G \otimes C(T)$ since $(Az)^t = A^0 z^t \in G$. If $u \in G \otimes C(T)$ then $u^t \in G$ for all t. Hence

$$\|z^t - A^0 z^t\|_1 \le \|z^t - u^t\|_1.$$

Expressed otherwise,

$$\int |z(s,t) - (Az)(s,t)| \, ds \le \int |z(s,t) - u(s,t)| \, ds.$$

When both members of this inequality are integrated with respect to t, the result is $\|z - Az\|_1 \le \|z - u\|_1$. ∎

Assuming now that A^0 and B^0 are continuous L_1-proximity maps, we conclude that the same is true of their extensions, A and B. The L_1-version of the Diliberto-Straus Algorithm then reads:

$$z_0 = z, \quad z_{k+1} = z_k - w_k, \quad w_k = Az_k \ (k \text{ odd}), \quad w_k = Bz_k \ (k \text{ even}).$$

We shall also require that $A^0(x + g) = A^0 x + g$ and $B^0(y + h) = B^0 y + h$ whenever $g \in G$ and $h \in H$.

7.2 LEMMA. If $u \in G \otimes L_1(T)$, $v \in L_1(S) \otimes H$, and $u + v \in C(S \times T)$, then $u \in G \otimes C(T)$ and $v \in C(S) \otimes H$.

PROOF. It suffices to prove that v is continuous. By 11.11 we can select biorthonormal bases, $\{g_i, \phi_i\}_1^n$ for $(G, \| \ \|_1)$ and $\{h_i, \psi_i\}_1^m$ for $(H, \| \ \|_1)$. Write $v = \sum_{i=1}^m x_i h_i$ and $u = \sum_{i=1}^n g_i y_i$, with $x_i \in L_1(S)$ and $y_i \in L_1(T)$. Put $w = u + v$. Then

$$\|y_i\|_1 = \int |y_i(t)| \, dt = \int |\langle \phi_i, u^t \rangle| \, dt \le \int \|u^t\|_1 \, dt = \|u\|_1.$$

Hence

$$\|u_s - u_\sigma\|_1 = \left\| \sum_{i=1}^n [g_i(s) - g_i(\sigma)] y_i \right\|_1 \le \sum_{i=1}^n |g_i(s) - g_i(\sigma)| \, \|y_i\|_1$$

$$\le \|u\|_1 \sum_{i=1}^n |g_i(s) - g_i(\sigma)|.$$

Now

$$|x_i(s) - x_i(\sigma)| = |\langle \psi_i, v_s - v_\sigma \rangle| \le \|v_s - v_\sigma\|_1 = \|w_s - u_s - w_\sigma + u_\sigma\|_1$$

$$\le \|w_s - w_\sigma\|_1 + \|u_s - u_\sigma\|_1$$

$$\le \|w_s - w_\sigma\|_\infty + \|u\|_1 \sum_{i=1}^n |g_i(s) - g_i(\sigma)|.$$

If $\sigma \to s$ then $\|w_s - w_\sigma\|_\infty \to 0$ by 11.7. Hence the preceding inequality establishes the continuity of x_i. Therefore v is continuous. ∎

7.3 LEMMA. *There exists a constant c such that each element w of $G \otimes C(T) + C(S) \otimes H$ has a representation $w = u + v$, with*

$$u \in G \otimes C(T), \quad v \in C(S) \otimes H, \text{ and } \|u\|_1 + \|v\|_1 \le c\|w\|_1.$$

PROOF. By 1.16 and 11.2, $G \otimes L_1(T) + L_1(S) \otimes H$ is closed (with respect to the L_1-norm) in $L_1(S \times T)$. By 11.3 there is a constant c such that each element w in $G \otimes L_1(T) + L_1(S) \otimes H$ has a representation $w = u + v$ with

$$u \in G \otimes L_1(T), \quad v \in L_1(S) \otimes H, \text{ and } \|u\|_1 + \|v\|_1 \le c\|w\|_1.$$

All that remains to be proved is that if $w \in G \otimes C(T) + C(S) \otimes H$, then u and v will be continuous. This is established by the preceding lemma. ∎

From now on, we write $U = G \otimes C(T)$, $V = C(S) \otimes H$, and $W = U + V$.

7.4 LEMMA. *Let $\{z_k\}_0^\infty$ be a sequence generated by the L_1-version of the Diliberto-Straus Algorithm. For a fixed k, suppose that $z - z_k = u + v$, with $u \in U$, $v \in V$, and $\|u\|_1 + \|v\|_1 \le \lambda\|z\|_1$. If k is even then*

$$\|u\|_\infty \le \lambda M \|z\|_\infty \qquad \left(M = 4 \sum \|h_i\|_\infty \sum \|g_i\|_\infty \right).$$

If k is odd, then v obeys this inequality.

PROOF. We give the proof for even k. The other case is deduced by considerations of symmetry. Let $\{g_i, \phi_i\}$ and $\{h_i, \psi_i\}$ be biorthonormal systems as in the proof of 7.2. Let $u = \sum g_i y_i$ and $v = \sum x_i h_i$. Then

$$\|x_i\|_1 = \int |x_i(s)| \, ds = \int |\langle \psi_i, v_s \rangle| \, ds \le \int \|v_s\|_1 \, ds = \|v\|_1 \le \lambda\|z\|_1 \le \lambda\|z\|_\infty.$$

Hence

$$\|v^t\|_1 = \left\| \sum h_i(t) x_i \right\|_1 \le \sum |h_i(t)| \|x_i\|_1 \le \lambda\|z\|_\infty \sum \|h_i\|_\infty.$$

Since k is even,

$$0 = A(z_{k-1} - Az_{k-1}) = A(z_k) = A(z - u - v) = A(z - v) - u.$$

Therefore $u = A(z - v)$ and $u^t = A^0(z^t - v^t)$. Thus

$$|y_i(t)| = |\langle \phi_i, u^t \rangle| \le \|u^t\|_1 \le 2\|z^t - v^t\|_1 \le 2\|z^t\|_1 + 2\|v^t\|_1$$
$$\le 2\|z\|_\infty + 2\lambda\|z\|_\infty \sum \|h_i\|_\infty.$$

Finally,

$$\|u\|_\infty = \left\| \sum g_i y_i \right\|_\infty \leq \sum \|g_i\|_\infty \|y_i\|_\infty$$

$$\leq \left\{ 2\|z\|_\infty + 2\lambda\|z\|_\infty \sum \|h_i\|_\infty \right\} \sum \|g_i\|_\infty$$

$$= 2\|z\|_\infty \left\{ 1 + \lambda \sum \|h_i\|_\infty \right\} \sum \|g_i\|_\infty$$

$$\leq 2\|z\|_\infty \left\{ 2\lambda \sum \|h_i\|_\infty \right\} \sum \|g_i\|_\infty = \lambda M \|z\|_\infty. \quad \blacksquare$$

7.5 LEMMA. *Let $\{z_k\}_0^\infty$ be a sequence generated by the L_1-version of the Diliberto-Straus Algorithm. Let c be the constant referred to in 7.3. Then for all k,*

$$\|z_k\|_\infty \leq 9cM\|z\|_\infty \qquad \left(M = 4 \sum \|h_i\|_\infty \sum \|g_i\|_\infty \right).$$

PROOF. Define $w_k, u_k,$ and v_k as in Chapter 5 (just after item 5.4). Then $u_k \in U$, $v_k \in V$, $z - z_{2k+1} = u_k + v_k$, and $z - z_{2k} = u_k + v_{k-1}$. Let k be fixed. By 7.3, there exists an element $d \in U \cap V$ such that

$$\|u_k - d\|_1 + \|v_{k-1} + d\|_1 \leq c\|z - z_{2k}\|_1 \leq 2c\|z\|_1.$$

We conclude from this and 7.4 that

$$\|u_k - d\|_\infty \leq 2cM\|z\|_\infty.$$

Now

$$z - z_{2k+1} = u_k + v_k = (u_k - d) + (v_k + d)$$

and

$$\|u_k - d\|_1 + \|v_k + d\|_1 = \|u_k - d\|_1 + \|v_k + u_k - (u_k - d)\|_1$$

$$\leq 2c\|z\|_1 + \|v_k + u_k\|_1 + \|u_k - d\|_1$$

$$\leq 2c\|z\|_1 + \|z - z_{2k+1}\|_1 + 2c\|z\|_1$$

$$\leq 4c\|z\|_1 + \|z\|_1 + \|z_{2k+1}\|_1$$

$$\leq (4c + 2)\|z\|_1 \leq 6c\|z\|_1.$$

Another application of 7.4 yields

$$\|v_k + d\|_\infty \leq 6cM\|z\|_\infty.$$

It follows that

$$\|z - z_{2k+1}\|_\infty \leq 8cM\|z\|_\infty \text{ and } \|z_{2k+1}\|_\infty \leq 9cM\|z\|_\infty.$$

The proof for z_{2k} is similar. $\quad \blacksquare$

At this point, we review the assumptions made about the subspaces G and H. These are finite-dimensional and possess L_1-proximity maps A^0 and B^0 respectively. We assume that A^0 and B^0 are continuous in the supremum norm and satisfy

$$A^0(x + g) = A^0 x + g \text{ and } B^0(y + h) = B^0 y + h$$

whenever $x \in C(S)$, $y \in C(T)$, $g \in G$, and $h \in H$.

78

7.6 THEOREM. *If, in addition to the assumptions of the preceding paragraph, it is hypothesized that B^0 is Lipschitz (in the supremum norm), then each sequence produced by the Diliberto-Straus Algorithm with the extended maps A and B is contained in a compact set.*

PROOF. Fix z in $C(S \times T)$ and let $\{z_k\}$ be the resulting sequence. Define A', B', and K by

$$A'f = A(z - f)$$
$$B'f = B(z - f)$$
$$Kf = f + B'f + A'(f + B'f).$$

An easy calculation shows that $K(z - z_{2k}) = z - z_{2k+2}$. Furthermore, if $u \in U$ and $v \in V$, then $K(u + v) = B'u + A'B'u$.

Since $U + V$ is closed, there exists (by 11.3) a constant c_0 such that each element $z - z_{2k}$ has a representation

$$z - z_{2k} = a_k + b_k$$

in which

$$a_k \in U, \quad b_k \in V, \quad \text{and} \quad \|a_k\|_\infty + \|b_k\|_\infty \le c_0 \|z - z_{2k}\|_\infty.$$

By the preceding lemma, the sequence $\|z_k\|_\infty$ is bounded, and so is the sequence $\|a_k\|_\infty$. Since B^0 is Lipschitz, B' is compact as an operator from U to V by 2.23. Hence $\{B'a_k\}$ lies in a compact subset of V. Since

$$z - z_{2k+2} = K(z - z_{2k}) = K(a_k + b_k)$$
$$= B'a_k + A'B'a_k,$$

and since A' is continuous, we see that $\{z - z_{2k}\}$ lies in a compact subset of $U + V$. From the equation

$$z_{2k+1} = z_{2k} - Bz_{2k}$$

we conclude that the sequence $\{z_{2k+1}\}$ also lies in a compact set. Hence $\{z_k\}$ lies in a compact set. ∎

The next two results hold for measure spaces without topology.

7.7 LEMMA. *Let $(S, \mathcal{A}, \mu)$ be a finite measure space. Let f be a measurable function, positive almost everywhere. Let $A_1, A_2, \ldots$ be measurable sets such that $\int_{A_k} f \to 0$. Then $\mu(A_k) \to 0$.*

PROOF. If the conclusion is false, we can assume (by passing to subsequences) that $\mu(A_k) \ge \delta > 0$ and that $\int_{A_k} f \le 2^{-k}$. Define $B_k = \cup_{i=k}^\infty A_i$ and $B = \cap_{k=1}^\infty B_k$. Since

79

$\mu(B_k) \geq \mu(A_k) \geq \delta$, since the B_k are nested, and since $\mu(B_1) < \infty$, we conclude [148, p. 17] that $\mu(B) \geq \delta$. Since

$$\int_B f \leq \int_{B_k} f \leq \sum_{i=k}^{\infty} \int_{A_i} f \leq \sum_{i=k}^{\infty} 2^{-i} = 2^{-k+1}$$

we conclude that $\int_B f = 0$. Hence f vanishes almost everywhere on B, contradicting the hypotheses. ∎

7.8 THEOREM. *Let S be a measure space, and G a linear subspace of $L_1(S)$. For $x \in L_1(S)$ these properties are equivalent:*

(i) $\|x\|_1 = \mathrm{dist}_1(x, S)$ *(i.e., 0 is a best approximation to x);*

(ii) $\int g(s)\mathrm{sgn}\, x(s)\, ds \leq \int |g(s)|\tilde{x}(s)\, ds$ *for all $g \in G$.*

Here $\tilde{x}$ denotes the characteristic function of $\{s : x(s) = 0\}$.

PROOF. This is a theorem of R.C. James. See [107] or [160, p. 46] for the proof. ∎

7.9 LEMMA. *Assume that z differs almost everywhere from each member of W. Let r_k be the characteristic function of $\{(s, t) : \mathrm{sgn}\, z_k(s, t) \neq \mathrm{sgn}\, z_{k+1}(s, t)\}$. Then $\iint r_k|z_k| \to 0$.*

PROOF. Recall that $\|z_k\|_1$ converges downward. Hence $\|z_k\|_1 - \|z_{k+1}\|_1 \to 0$. We shall prove that

(1) $$2 \iint r_k|z_k| = \|z_k\|_1 - \|z_{k+1}\|_1.$$

Suppose that k is even. Then $z_{k+1} = z_k - Bz_k$. By the characterization theorem (7.8),

$$\iint v\, \mathrm{sgn}\, z_{k+1}\, d\sigma = 0 \qquad (v \in V).$$

Taking $v = xh$, for arbitrary $x \in C(S)$ and $h \in H$, we have

$$\iint x(s)h(t)\, \mathrm{sgn}\, z_{k+1}(s, t)\, ds\, dt = 0.$$

Since x is arbitrary in $C(S)$ we conclude from 11.10 and the Fubini Theorem that

$$\int h(t)\, \mathrm{sgn}\, z_{k+1}(s, t)\, dt = 0 \qquad (h \in H, \ s \in S).$$

Since $z_{k+1} - z_k \in V$, $z_{k+1}(s, \cdot) - z_k(s, \cdot) \in H$. Hence

$$\int [z_{k+1}(s, t) - z_k(s, t)]\, \mathrm{sgn}\, z_{k+1}(s, t)\, dt = 0 \qquad (s \in S).$$

Integrating over S yields

$$\iint (z_{k+1} - z_k)\, \text{sgn}\, z_{k+1} = 0.$$

This gives us

$$\|z_{k+1}\|_1 = \iint z_k\, \text{sgn}\, z_{k+1}$$

$$= \iint z_k\, \text{sgn}\, z_k - \iint z_k(\text{sgn}\, z_k - \text{sgn}\, z_{k+1})$$

$$= \|z_k\|_1 - \iint r_k z_k(\text{sgn}\, z_k - \text{sgn}\, z_{k+1})$$

$$= \|z_k\|_1 - \iint r_k z_k(\text{sgn}\, z_k + \text{sgn}\, z_k).$$

The last step uses the fact that z_k and z_{k+1} are almost everywhere different from 0. The above equation is a rearrangement of Eq. (1). The case when k is odd is the same, *mutatis mutandis.* ∎

7.10 LEMMA. *Assume that z differs almost everywhere from each member of W. If $\{z_i; i \in I\}$ is a uniformly convergent subsequence of $\{z_k\}$, then in the notation of 7.9*

$$\lim_{i \in I} \iint r_i\, d\sigma = 0.$$

PROOF. Suppose that $\|f - z_i\|_\infty \to 0$. From the inequality

$$\iint r_i |f| \leq \iint r_i |f - z_i| + \iint r_i |z_i|$$

and from the preceding lemma, we draw the conclusion that

$$\lim_{i \in I} \iint r_i |f| = 0.$$

Since W is closed and $z - z_i \in W$, we have $z - f \in W$. Thus $f = z - w$ for some $w \in W$, and consequently the zero set of f is a null set. Lemma 7.7 therefore applies to draw the conclusion $\iint r_i \to 0$. ∎

7.11 LEMMA. *Let $\{x_n\}$ be a sequence in $C(S)$ which converges uniformly to a function x. If $\mu\{s : x(s) = 0\} = 0$, then $\text{sgn}\, x_n$ converges in the metric of L_1 to $\text{sgn}\, x$.*

PROOF. First we shall prove that $\mu(F_n) \to 0$, where

$$F_n = \{s : \text{sgn}\, x_n(s) \neq \text{sgn}\, x(s)\}.$$

After that, the proof is completed by writing

$$\int |\text{sgn}\, x_n - \text{sgn}\, x| \leq 2\mu(F_n) \to 0.$$

If $\mu(F_n) \nrightarrow 0$, then for a suitable $\delta > 0$ and a suitable infinite set J of integers we have, for $j \in J$, $\mu(F_j) \geq \delta$. Define $A_n = \{s : |x(s)| < 1/n\}$. We assert that $\mu(A_n) \geq \delta$, for all n. To verify this, fix n and select $j \in J$ so that $\|x_j - x\|_\infty < 1/n$. Then $F_j \subset A_n$. Indeed, if $s \notin A_n$ then either $x(s) \geq 1/n$ or $x(s) \leq -1/n$. In the first case, $x_j(s) > x(s) - 1/n \geq 1/n - 1/n = 0$. In the second case, $x_j(s) < x(s) + 1/n \leq -1/n + 1/n = 0$. In either case, $\operatorname{sgn} x_j(s) = \operatorname{sgn} x(s)$, and $s \notin F_j$. Thus we have $\mu(A_n) \geq \mu(F_j) \geq \delta$. Now $A_1 \supset A_2 \supset \cdots$ and $\mu(A_1) < \infty$. Hence [**148**, p. 17] we may conclude that

$$\delta \leq \lim \mu(A_n) = \mu\left(\bigcap_{n=1}^{\infty} A_n\right) = \mu\{s : x(s) = 0\} = 0.$$

This contradiction completes the proof. ∎

7.12 THEOREM. *Assume the hypotheses of Theorem 7.6. Let z be an element of $C(S \times T)$ which differs almost everywhere from each element of W. Then the L_1-version of the Diliberto-Straus Algorithm produces from z a sequence $\{z_k\}$ such that $\|z_k\|_1 \downarrow \operatorname{dist}_1(z, W)$. Furthermore, the sequence $\{z - z_n\}$ possesses cluster points, and each of them is a best L_1-approximation to z in the subspace W.*

PROOF. By 7.6, $\{z_k\}$ lies in a compact subset of $C(S \times T)$. Let f be any cluster point of this sequence, and let $\|z_{k_i} - f\|_\infty \to 0$ as $i \to \infty$. We shall prove that $z - f \in W$ and that $z - f$ is a best approximation to z in the L_1-metric. Since $z - z_k \in W$ and W is closed, it is clear that $z - f \in W$. In order to establish the best approximation property, we use the characterization theorem (7.8). Thus it suffices to prove that

$$\iint w \operatorname{sgn} f = 0 \qquad (w \in W).$$

For this, it suffices to prove that

$$\iint u \operatorname{sgn} f = 0 \text{ and } \iint v \operatorname{sgn} f = 0 \qquad (v \in V,\, u \in U).$$

These two proofs are similar, and we consider only the first equation. Since $u(s, t)$ is a sum of terms $y(t)g(s)$ with $y \in C(T)$ and $g \in G$ it suffices to prove

$$\int g(s) \operatorname{sgn} f(s, t)\, ds = 0 \qquad (g \in G,\, t \in T).$$

Fix $g \in G$. If k is even, then $z_k = z_{k-1} - Az_{k-1}$, and so by 7.8,

$$\iint u \operatorname{sgn} z_k = 0.$$

An application of 11.10 yields

$$\int g \operatorname{sgn} z_k\, ds = 0$$

82

for almost all $t \in T$. If k is odd, then

$$\int g \operatorname{sgn} z_k \, ds = \int g \operatorname{sgn} z_{k+1} \, ds + \int g(\operatorname{sgn} z_k - \operatorname{sgn} z_{k+1}) \, ds$$

$$= \int g r_k (\operatorname{sgn} z_k - \operatorname{sgn} z_{k+1}) \, ds$$

where r_k is as in 7.9. It follows that

$$\int \left| \int g \operatorname{sgn} z_k \, ds \right| dt = \int \left| \int g r_k (\operatorname{sgn} z_k - \operatorname{sgn} z_{k+1}) \, ds \right| dt$$

$$= \int \left| \int 2 g r_k \operatorname{sgn} z_k \, ds \right| dt$$

$$\le 2 \int\int |g| r_k \, ds \, dt$$

$$\le 2 \|g\|_\infty \int\int r_k .$$

This inequality is therefore true whether k is even or odd, and so

$$\int \left| \int g \operatorname{sgn} f \, ds \right| dt \le \int \left| \int g(\operatorname{sgn} f - \operatorname{sgn} z_k) \, ds \right| dt + \int \left| \int g \operatorname{sgn} z_k \, ds \right| dt$$

$$\le \|g\|_\infty \int\int |\operatorname{sgn} f - \operatorname{sgn} z_k| + 2\|g\|_\infty \int\int r_k.$$

By the preceding lemma, $\int\int |\operatorname{sgn} f - \operatorname{sgn} z_{k_i}| \to 0$. By 7.10, $\int\int r_{k_i} \to 0$. Hence

$$\int \left| \int g \operatorname{sgn} f \, ds \right| dt = 0 \qquad \text{and} \qquad \int g \operatorname{sgn} f \, ds = 0$$

for almost all t. ∎

7.13 EXAMPLE. We exhibit a function z in $C(S \times T)$ for which the L_1-version of the Diliberto-Straus Algorithm fails, in the sense that $\lim \|z_k\|_1 > \operatorname{dist}_1(z, W)$. Here $W = C(S) + C(T)$. We let $S = T = [-1, 1]$ and define

$$z(s, t) = \begin{cases} st & \text{if } s > 0 \text{ and } t > 0; \\ -st & \text{if } s < 0 \text{ and } t < 0; \\ 0 & \text{otherwise.} \end{cases}$$

With the help of 7.8 it is easily seen that $Bz = Az = 0$, and so $z_k = z$ for all k. However, 0 is not a best approximation to z, since, again by 7.8, and easy calculations,

$$\int\int w \operatorname{sgn} z > \int\int |w| \tilde{z}$$

if we take $w(s, t) = s + t$.

7.14 LEMMA. Let H be a one-dimensional subspace of $C(T)$ generated by a positive function, h. For $x \in C(T)$ these are equivalent properties:

(i) $\|x\|_1 = \text{dist}_1(x, H)$

(ii) $\int_{P(x)} h \leq \frac{1}{2} \int h$ and $\int_{N(x)} h \leq \frac{1}{2} \int h$

where $P(x) = \{t : x(t) > 0\}$ and $N(x) = \{t : x(t) < 0\}$.

PROOF. By James' characterization theorem, (7.8), (i) is equivalent to

$$(2) \qquad \left| \int \lambda h \, \text{sgn} \, x \right| \leq \int_{Z(x)} |\lambda h| \qquad (\lambda \in \mathbb{R}).$$

Here $Z(x)$ is the set where $x(t) = 0$. Then because $h > 0$, (2) is equivalent to

$$(3) \qquad \left| \int h \, \text{sgn} \, x \right| \leq \int_{Z(x)} h.$$

Further equivalences are:

$$(4) \qquad \left| \int_P h \, \text{sgn} \, x + \int_N h \, \text{sgn} \, x \right| \leq \int_Z h$$

$$(5) \qquad \left| \int_P h - \int_N h \right| \leq \int_Z h = \int h - \int_N h - \int_P h$$

$$(6) \qquad \int_P h - \int_N h \leq \int h - \int_N h - \int_P h \quad \text{and} \quad \int_N h - \int_P h \leq \int h - \int_N h - \int_P h$$

$$(7) \qquad 2 \int_P h \leq \int h \quad \text{and} \quad 2 \int_N h \leq \int h. \quad \blacksquare$$

7.15 LEMMA. Let H be a one-dimensional subspace of $C(T)$ generated by a positive function h. Then there exists an L_1-proximity map $B : C(T) \twoheadrightarrow H$ such that

(i) B is monotone: $Bx \geq By$ if $x \geq y$.

(ii) $B(y + \beta h) = By + \beta h$ for $y \in C(T)$ and $\beta \in \mathbb{R}$.

(iii) B is Lipschitz in the supremum norm.

PROOF. Define a nonlinear functional $\lambda : C(T) \to \mathbb{R}$ by

$$\lambda(y) = \sup\{\alpha \in \mathbb{R} : \|y - \alpha h\|_1 = \text{dist}_1(y, H)\}.$$

Then define B by $By = \lambda(y)h$. Since the set of best L_1-approximations to y in H is a compact convex set, $\lambda(y)h$ is one of the best approximations of y.

In order to prove (i), let $x, y \in C(T)$ with $x \geq y$. It suffices to show that $\lambda(x) \geq \lambda(y)$. If $\lambda(x) < \lambda(y)$, then $\lambda(y)h$ is not a best approximation to x. By 7.14, either

$$\int_{P(x-\lambda(y)h)} h > \tfrac{1}{2} \int h \quad \text{or} \quad \int_{N(x-\lambda(y)h)} h > \tfrac{1}{2} \int h.$$

Since $x \geq y$,

$$N(x - \lambda(y)h) \subset N(y - \lambda(y)h).$$

Since $\lambda(x) < \lambda(y)$,

$$P(x - \lambda(y)h) \subset P(x - \lambda(x)h).$$

Hence either

$$\int_{P(x-\lambda(x)h)} h > \tfrac{1}{2} \int h \quad \text{or} \quad \int_{N(y-\lambda(y)h)} h > \tfrac{1}{2} \int h.$$

By 7.14 again, either $\lambda(x)h$ is not a best approximation of x or $\lambda(y)h$ is not a best approximation of y.

For part (ii), we note that if $\|x - \alpha h\|_1 = \mathrm{dist}_1(x, H)$, then

$$\|x + \beta h - (\alpha + \beta)h\|_1 = \mathrm{dist}_1(x, H) = \mathrm{dist}_1(x + \beta h, H).$$

In other words,

$$\beta + \{\alpha : \|x - \alpha h\|_1 = \mathrm{dist}_1(x, H)\} \subset \{\alpha : \|x + \beta h - \alpha h\|_1 = \mathrm{dist}_1(x + \beta h, H)\}.$$

It follows that $\beta + \lambda(x) \leq \lambda(x + \beta h)$. If we replace β by $-\beta$ and then x by $x + \beta h$ we get $\lambda(x + \beta h) \leq \lambda(x) + \beta$. Hence

$$\lambda(x + \beta h) = \lambda(x) + \beta \quad \text{and} \quad B(x + \beta h) = Bx + \beta h.$$

For part (iii), we assume without loss of generality that $h \geq 1$. For any x and y in $C(T)$ we have

$$|x - y| \leq \|x - y\|_\infty h.$$

Equivalently,

$$-\|x - y\|_\infty h + y \leq x \leq \|x - y\|_\infty h + y.$$

Using parts (i) and (ii) above, we have

$$-\|x - y\|_\infty h + By \leq Bx \leq \|x - y\|_\infty h + By.$$

85

Equivalently,

$$-\|x - y\|_\infty h \le Bx - By \le \|x - y\|_\infty h$$

and

$$\|Bx - By\|_\infty \le \|x - y\|_\infty \|h\|_\infty. \quad \blacksquare$$

7.16 THEOREM. *Let S and T be compact intervals on the real line, each with Lebesgue measure. Let G be a finite-dimensional Haar subspace of $C(S)$, and let H be a one-dimensional Haar subspace of $C(T)$. Let z be an element of $C(S \times T)$ which differs almost everywhere from each element of $W = C(S) \otimes H + G \otimes C(T)$. Then the L_1-version of the Diliberto-Straus algorithm produces a sequence $\{z_k\}$ which possesses cluster points, and $\|z_k\|_1 \downarrow \mathrm{dist}_1(z, W)$.*

PROOF. By a theorem of Jackson [**160**, p. 236], each element of $C(S)$ has a unique best L_1-approximation in G, and each element of $C(T)$ has a unique best L_1-approximation in H. Hence G and H have uniquely-determined L_1-proximity maps, A^0 and B^0. It follows that

$$A^0(x + g) = A^0 x + g \qquad (x \in C(S), \ g \in G)$$

and that B^0 has the analogous property. The L_1-continuity of A^0 is guaranteed by a theorem from [**104**, p. 164]. Now if $\{x_n\}$ is a sequence in $C(S)$ converging uniformly to x, then $\{x_n\}$ converges to x in the L_1-norm, and so $\{A^0 x_n\}$ converges to $A^0 x$ in the L_1-norm. Since the supremum norm and the L_1-norm are equivalent on the finite-dimensional space G, $\{A^0 x_n\}$ converges uniformly to $A^0 x$. Hence A^0 is continuous in the supremum norm. Since H is a one-dimensional Haar subspace and T is an interval, H is generated by a positive function. By 7.15, B^0 is a Lipschitz map. The conclusions of the theorem now follow from 7.12. $\blacksquare$

7.17 LEMMA. *Let H be a finite-dimensional subspace of $C(T)$ such that each element of $C(T)$ has a unique best L_1-approximation in H. Let $B^0 : C(T) \twoheadrightarrow H$ be the L_1-proximity map, and B its extension to $C(S \times T)$. Let $f_k \in C(S \times T)$ and $f_k \to f$ uniformly. If f is different from 0 almost everywhere and if*

$$\int \left| \int h(t) \, \mathrm{sgn} \, f_k(s,t) \, dt \right| ds \to 0 \qquad (h \in H)$$

then $Bf = 0$.

PROOF. By 7.11,

$$\iint |\mathrm{sgn} \, f - \mathrm{sgn} \, f_k| \, d\sigma \to 0.$$

Therefore

$$\int \left| \int h \, \text{sgn} \, f \, dt \right| ds \leq \int \left| \int h(\text{sgn} \, f - \text{sgn} \, f_k) \, dt \right| ds + \int \left| \int h \, \text{sgn} \, f_k \, dt \right| ds \rightarrow 0.$$

This shows that $\int h(t) \, \text{sgn} \, f(s,t) \, dt = 0$ for almost all s and for all $h \in H$. For all $x \in C(S)$ we therefore have

$$\iint x(s) h(t) \, \text{sgn} \, f(s,t) \, dt \, ds = 0.$$

By the Characterization Theorem (7.8), 0 is a best L_1-approximation of f from the subspace $V = C(S) \otimes H$. Since best L_1-approximations in H are unique, the same is true of V, by 2.11. Hence 0 is the best L_1-approximation of f in V, and consequently, $Bf = 0$. ∎

7.18 LEMMA. *In the L_1-version of the Diliberto-Straus Algorithm assume that A^0 and B^0 are nonexpansive in the supremum norm. Then*

$$\|z_1 - z_0\|_\infty \geq \|z_2 - z_1\|_\infty \geq \|z_3 - z_2\|_\infty \geq \cdots.$$

PROOF. Observe first that the extension A is also nonexpansive since

$$\begin{aligned}
\|Af_1 - Af_2\|_\infty &= \sup_t \sup_s |(Af_1)(s,t) - (Af_2)(s,t)| \\
&= \sup_t \sup_s |(A^0 f_1^t)(s) - (A^0 f_2^t)(s)| \\
&= \sup_t \|A^0 f_1^t - A^0 f_2^t\|_\infty \\
&\leq \sup_t \|f_1^t - f_2^t\|_\infty \\
&= \|f_1 - f_2\|_\infty.
\end{aligned}$$

Now let k be an odd integer. Then

$$\|z_{k+1} - z_k\| = \|Az_k\| = \|Az_k - Az_{k-1}\| \leq \|z_k - z_{k-1}\|.$$

The proof for even k is similar. ∎

7.19 LEMMA. *Assume the hypotheses of 7.12, and assume that A^0 and B^0 are nonexpansive in the supremum norm. Assume also that best L_1-approximations in G and H are unique. If z is an element of $C(S \times T)$ which differs almost everywhere from each member of W, then in the algorithm, $\|z_{k+1} - z_k\|_\infty \rightarrow 0$.*

PROOF. By 7.6, the sequence $\{z_k\}$ is bounded and equicontinuous. Let f be a cluster point of the sequence, and let $z_{k_i} \rightarrow f$. By 7.12, $z - f$ is a best L_1-approximation of z in W, and so 0 is a best L_1-approximation of f in W. Consequently 0 is a best L_1-approximation of f in U and in V. Since best L_1-approximations in U and V are unique, it follows from

2.11 that $Af = Bf = 0$. By the fact that A is nonexpansive (see the preceding proof) we have

$$\|Az_{k_i}\|_\infty = \|Az_{k_i} - Af\|_\infty \leq \|z_{k_i} - f\|_\infty \to 0.$$

Similarly $Bz_{k_i} \to 0$. Now by the preceding lemma,

$$\|Bz_0\|_\infty \geq \|Az_1\|_\infty \geq \|Bz_2\|_\infty \geq \|Az_3\|_\infty \geq \cdots$$

and so it follows that $Bz_k \to 0$ and $Az_k \to 0$. Since $z_{k+1} - z_k$ is either Az_k or Bz_k, we conclude that $\|z_{k+1} - z_k\|_\infty \to 0$. ∎

7.20 THEOREM. *Let G and H be the one-dimensional spaces of constant functions in $C(S)$ and $C(T)$ respectively, so that $W = C(T) + C(S)$. If $z \in C(S \times T)$ and differs almost everywhere from each element of W, then the sequence $\{z_n\}$ generated by the L_1-version of the Diliberto-Straus Algorithm converges uniformly, and $z - \lim z_n$ is a best L_1-approximation of z in W.*

PROOF. The L_1-proximity map $B^0 : C(T) \twoheadrightarrow H$ described in 7.15 is Lipschitz. An examination of the proof of 7.15 shows that in fact B^0 is nonexpansive. (The Lipschitz constant in 7.15 is $\|h\|_\infty$.) Similarly there is a nonexpansive L_1-proximity map $A^0 : C(S) \twoheadrightarrow G$.

We define A and B by the usual extensions of A^0 and B^0. The extended maps, defined on $C(S \times T)$, are also nonexpansive. (See the proof of 7.17). Fixing $z \in C(S \times T)$, we put $A'w = A(z - w)$ and $B'w = B(z - w)$, where $w \in W$. Define u_k and v_k as in Chapter 5, just before 5.5. Then we have $u_{k+1} = A'B'u_k$. As in 5.5, $|u_k(s) - u_k(\sigma)| \leq \|z_s - z_\sigma\|_\infty$, so that $\{u_k\}$ is equicontinuous. Let f be any cluster point of $\{z_k\}$. Then as in 7.19, $Af = Bf = 0$. Since W is closed, $f = z - u - v$ for appropriate $u \in U$ and $v \in V$. The calculation in 5.7 shows that $\{u_k\}$ is bounded.

Now we know that $\{u_k\}$ is bounded and equicontinuous. By the Ascoli Theorem, this sequence contains a (uniformly) convergent subsequence, say $u_{k_i} \to u^*$. Since $A'B'$ is continuous and $u_{k+1} = A'B'u_k$, we have $u_{k_i+1} \to A'B'u^*$. By 7.19, $z_{k+1} - z_k \to 0$. Since

(8)
$$z_{2k} = z - u_k - v_{k-1} \quad \text{and} \quad z_{2k+1} = z - u_k - v_k,$$

we have

$$v_k - v_{k-1} = z_{2k} - z_{2k+1} \to 0.$$

Similarly, $u_k - u_{k-1} \to 0$. Therefore $u_{k_i+1} \to u^*$, and $u^* = A'B'u^*$. Then we have

$$\|u_{k+1} - u^*\|_\infty = \|A'B'u_k - A'B'u^*\|_\infty \leq \|u_k - u^*\|_\infty.$$

This shows that the sequence $\{u_k\}$ itself converges to u^*. Since $v_k = B'u_k$, $v_k \to B'u^*$. Using (8), we conclude that $z - z_k \to u^* + B'u^*$. By 7.12, $\|z - u^* - B'u^*\|_1 = \text{dist}_1(z, W)$. ∎

7.21 LEMMA. *The constant M in 7.4 has the property*

$$\|A(z-v)\|_\infty \le M(\|z\|_\infty + \|v\|_1) \qquad z \in C(S \times T), \ v \in V.$$

PROOF. Put $u = A(z-v)$. Let

$$u(s,t) = \sum_{i=1}^{n} g_i(s)y_i(t) \ \text{ and } \ v(s,t) = \sum_{j=1}^{m} x_j(s)h_j(t).$$

As in the proof of 7.4 we have $\|x_i\|_1 \le \|v\|_1$. Hence

$$\|v^t\|_1 = \|\sum_{j=1}^{m} h_j(t)x_j\|_1 \le \max_i \|x_i\|_1 \sum_{j=1}^{m} |h_j(t)| \le \|v\|_1 \sum_{j=1}^{m} \|h_j\|_\infty.$$

Furthermore

$$\|u^t\|_1 = \|A_G(z^t - v^t)\|_1 \le 2\|z^t - v^t\|_1 \le 2\|z^t\|_1 + 2\|v^t\|_1$$
$$\le 2\|z^t\|_\infty + 2\|v^t\|_1 \le 2\|z\|_\infty + 2\|v^t\|_1$$
$$\le 2\|z\|_\infty + 2\|v\|_1 \sum_{j=1}^{m} \|h_j\|_\infty.$$

Again, as in the proof of 7.4,

$$|y_i(t)| \le \|u^t\|_1 \le 2\|z\|_\infty + 2\|v\|_1 \sum_{j=1}^{m} \|h_j\|_\infty.$$

Hence

$$\|u\|_\infty \le \sum_{i=1}^{n} \|g_i\|_\infty \|y_i\|_\infty$$
$$\le \left\{ 2\|z\|_\infty + 2\|v\|_1 \sum_{j=1}^{m} \|h_j\|_\infty \right\} \sum_{i=1}^{n} \|g_i\|_\infty$$
$$\le M(\|z\|_\infty + \|v\|_1). \ \blacksquare$$

7.22 THEOREM. *Let G be a finite-dimensional subspace of $C(S)$ having an L_1-proximity map which is L_∞-continuous. Let H be a finite-dimensional subspace of $C(T)$ having an L_1-proximity map which satisfies an L_∞-Lipschitz condition. Then the subspace*

$$W \equiv G \otimes C(T) + C(S) \otimes H$$

is L_1-proximinal in $C(S \times T)$.

PROOF. By 7.1, the subspace $V = C(S) \otimes H$ is L_1-proximinal. The proximity maps A^0, A, B^0, B are defined as at the beginning of this chapter. Fixing $z \in C(S \times T)$, we define $\Gamma : U \to V$ by the equation $\Gamma u = B(z - u)$. Then Γ is L_∞-compact by 2.23.

89

Select elements $w_k \in W$ so that

$$\lim_{k \to \infty} \|z - w_k\|_1 = \mathrm{dist}_1(z, W).$$

There is no loss of generality in supposing that

$$\|w_k\|_1 \leq \|w_k - z\|_1 + \|z\|_1 \leq 2\|z\|_1 \leq 2\|z\|_\infty.$$

By 7.3, we can express each w_k in the following form:

$$w_k = u_k + v_k \quad \text{where} \quad \|u_k\|_1 + \|v_k\|_1 \leq c\|w\|_1 \leq 2c\|z\|_\infty.$$

Now define $u'_k = A(z - v_k)$. Then

$$\|z - u'_k - v_k\|_1 = \|z - u_k - v_k - A(z - u_k - v_k)\|_1$$
$$\leq \|z - u_k - v_k\|_1 = \|z - w_k\|_1.$$

Furthermore, by 7.21 we have

$$\|u'_k\|_\infty \leq M(\|z\|_\infty + \|v_k\|_1) \leq (2c + 1)M\|z\|_\infty.$$

Now define $v'_k = B(z - u'_k) = \Gamma(u'_k)$. By the same argument as above,

$$\|z - u'_k - v'_k\|_1 \leq \|z - u'_k - v_k\|_1 \leq \|z - w_k\|_1.$$

Since Γ is compact and $\{u'_k\}$ is bounded, the sequence $\{v'_k\}$ lies in a compact set. By the continuity of A, the sequence of points $u''_k = A(z - v'_k)$ lies in a compact set. The sequence $\{u''_k + v'_k\}$ has cluster points, and each of them is a best L_1-approximation of z in W. The latter is closed, by 11.2. ∎

CHAPTER 8

ESTIMATES OF PROJECTION CONSTANTS

Let Z be a Banach space and W a complemented subspace of Z. The **relative projection constant** of W in Z is defined to be the real number

$$\lambda(W, Z) = \inf\{\|P\| : P \text{ is a projection from } Z \text{ onto } W\}.$$

If this infimum is *attained* then the projections of least norm are called minimal projections from Z onto W. In this chapter we are concerned with estimating $\lambda(W, Z)$ when Z is a tensor product of Banach spaces. We defer until the next chapter the problem of finding minimal projections. Some well-known results on projections may be found in Chapter 11.

8.1 THEOREM. *Let G and H be complemented subspaces in Banach spaces X and Y respectively. For any uniform α on $X \otimes Y$, $G \overline{\otimes} Y + X \overline{\otimes} H$ is complemented (and therefore closed) in $X \otimes_\alpha Y$. Its relative projection constant does not exceed*

$$\lambda(G, X) + \lambda(H, Y) + \lambda(G, X)\lambda(H, Y).$$

PROOF. Let $P : X \twoheadrightarrow G$ and $Q : Y \twoheadrightarrow H$ be projections. By 11.2, $(P \otimes_\alpha I) \oplus (I \otimes_\alpha Q)$ is a projection of $X \otimes_\alpha Y$ onto the subspace

$$W = G \overline{\otimes} Y + X \overline{\otimes} H.$$

By the uniformity of α

$$\|(P \otimes_\alpha I) \oplus (I \otimes_\alpha Q)\| \le \|P\| + \|Q\| + \|P\|\,\|Q\|.$$

By taking an infimum as P and Q range over all appropriate projections we obtain

$$\lambda(W, X \otimes_\alpha Y) \le \lambda(G, X) + \lambda(H, Y) + \lambda(G, X)\lambda(H, Y). \quad \blacksquare$$

8.2 LEMMA. *Let A be a rectangular matrix with the property that for some pair of integers (μ, ν) we have*

$$\sum_i a_{i\nu} = \sum_j a_{\mu j} = \sum_{i,j} a_{ij} = 1.$$

Then $\sum_{i,j} |a_{ij}| \geq 3 - 2a_{\mu\nu}$.

PROOF. Define

$$\epsilon_{ij} = \begin{cases} +1 & \text{if } i = \mu \text{ or } j = \nu \\ -1 & \text{if } i \neq \mu \text{ and } j \neq \nu. \end{cases}$$

One readily verifies that

$$\epsilon_{ij} = 2(\delta_{i\mu} + \delta_{j\nu} - \delta_{i\mu}\delta_{j\nu}) - 1.$$

Then

$$\sum_{i,j} |a_{ij}| \geq \sum_{i,j} \epsilon_{ij} a_{ij}$$

$$= \sum_{i,j} (2\delta_{i\mu} + 2\delta_{j\nu} - 2\delta_{i\mu}\delta_{j\nu} - 1)a_{ij}$$

$$= 2\sum_{j} a_{\mu j} + 2\sum_{i} a_{i\nu} - 2a_{\mu\nu} - \sum_{i,j} a_{ij}$$

$$= 2 + 2 - 2a_{\mu\nu} - 1$$

$$= 3 - 2a_{\mu\nu}. \blacksquare$$

8.3 LEMMA. *Let G and H be subspaces of dimensions n and m in Banach spaces X and Y respectively. Then G and H have bases of norm-1 elements $\{g_1, \ldots, g_n\}$ and $\{h_1, \ldots, h_m\}$ with the following property. For any uniform reasonable crossnorm α on $X \otimes Y$ and for any element w in $G \otimes Y + X \otimes H$ there is a representation*

$$w = \sum_{i=1}^{n} g_i \otimes y_i + \sum_{i=1}^{m} x_i \otimes h_i \text{ with } \sum_{i=1}^{m} \|x_i\| + \sum_{i=1}^{n} \|y_i\| \leq 3n^2 m^2 \alpha(w).$$

PROOF. Select a biorthonormal set $\{g_i, \varphi_i\}$ for G. Thus

$$g_i \in G, \quad \varphi_i \in X^*, \quad \|g_i\| = \|\varphi_i\| = 1, \quad \varphi_i(g_j) = \delta_{ij} \quad (1 \leq i, j \leq n).$$

Define a projection P from X onto G by the equation $Px = \sum_{i=1}^{n} \varphi_i(x)g_i$. Then $\|P\| \leq n$. Similarly, we construct a projection Q from Y onto H of the form $Qy = \sum_{i=1}^{m} \psi_i(y)h_i$ with $\|Q\| \leq m$. As in the proof of 11.2 the operator $\overline{P} = P \otimes_\alpha I$ is a projection of $X \otimes_\alpha Y$ onto $G \otimes Y$, and $\|\overline{P}\| \leq n$. Similar remarks apply to $\overline{Q} = I \otimes_\alpha Q$. By 11.2, the operator $\overline{P} + \overline{Q} - \overline{Q}\,\overline{P}$ is a projection of $X \otimes_\alpha Y$ onto the subspace

$$W = G \otimes Y + X \otimes H.$$

For any $w \in W$ we write

$$w = \overline{P}w + \overline{Q}(I - \overline{P})w.$$

This is the representation of w referred to in the statement of the theorem. In order to verify the upper bounds given, consider the element $u = \overline{P}w$. It is of the form $u = \sum_{i=1}^{n} g_i \otimes y_i$ for appropriate $y_i \in Y$. We have

$$\alpha(u) = \alpha(\overline{P}w) \leq \|\overline{P}\|\alpha(w) \leq n\alpha(w).$$

If $\psi \in Y^*$ and $\|\psi\| = 1$, then (because α is a reasonable norm),

$$\alpha^*(\varphi_i \otimes \psi) = \|\varphi_i\|\,\|\psi\| = 1.$$

Hence

$$(\varphi_i \otimes \psi)(u) \leq \alpha^*(\varphi_i \otimes \psi)\alpha(u) \leq n\alpha(w).$$

But

$$(\varphi_i \otimes \psi)(u) = \sum_{j=1}^{n} \varphi_i(g_j)\psi(y_j) = \psi(y_i).$$

Hence

$$\psi(y_i) \leq n\alpha(w).$$

Since ψ was arbitrary, we conclude that $\|y_i\| \leq n\alpha(w)$. A similar calculation with the element $v = \overline{Q}(w - u)$ shows that $v = \sum_{i=1}^{m} x_i \otimes h_i$ with

$$\|x_i\| \leq m(1 + n)\alpha(w).$$

It follows that

$$\sum_{i=1}^{m} \|x_i\| + \sum_{i=1}^{n} \|y_i\| \leq \left[m^2(1+n) + n^2\right]\alpha(w) \leq 3n^2m^2\alpha(w). \quad \blacksquare$$

8.4 LEMMA. *Let G be an n-dimensional subspace in a Banach space X, and let H be an m-dimensional subspace in a Banach space Y. Let $\{x_i, \varphi_i\}_{i=1}^{k}$ and $\{y_i, \psi_i\}_{i=1}^{\ell}$ be biorthonormal systems in X and Y respectively. Assume that*
  (i) $\sum_{\mu=1}^{k} |\varphi_\mu(g)| \leq c\|g\|$ $(g \in G)$;
  (ii) $\sum_{\nu=1}^{\ell} |\psi_\nu(h)| \leq c\|h\|$ $(h \in H)$.

Let α be a uniform reasonable crossnorm. If P is a projection of $X \otimes_\alpha Y$ onto $G \otimes Y + X \otimes H$, then

$$\sum_{\mu\nu} |(\varphi_\mu \otimes \psi_\nu)(P(x_\mu \otimes y_\nu))| \leq c(n\ell + mk)3n^2m^2\|P\|.$$

PROOF. By 8.3 there exist norm-1 bases $\{g_1, \ldots, g_n\}$ for G and $\{h_1, \ldots, h_m\}$ for H such that

$$P(x_\mu \otimes y_\nu) = \sum_{i=1}^{m} x_i^{\mu\nu} \otimes h_i + \sum_{i=1}^{n} g_i \otimes y_i^{\mu\nu}$$

93

and

$$\sum_{i=1}^{m} \|x_i^{\mu\nu}\| + \sum_{j=1}^{n} \|y_j^{\mu\nu}\| \le 3n^2 m^2 \, \alpha(P(x_\mu \otimes y_\nu)) \le 3n^2 m^2 \, \|P\|.$$

Now

$$\sum_{\mu\nu} |(\varphi_\mu \otimes \psi_\nu) \left(\sum_{i=1}^{m} x_i^{\mu\nu} \otimes h_i \right)| \le \sum_{i\mu\nu} |\varphi_\mu(x_i^{\mu\nu}) \psi_\nu(h_i)|$$

$$\le \sum_{i\mu\nu} \|\varphi_\mu\| \, \|x_i^{\mu\nu}\| \, |\psi_\nu(h_i)| \le 3n^2 m^2 \|P\| \sum_{i\mu\nu} |\psi_\nu(h_i)|$$

$$\le 3n^2 m^2 \|P\| \sum_{\mu i} c\|h_i\| = 3n^2 m^2 mkc\|P\|.$$

A similar calculation gives

$$\sum_{\mu\nu} |(\varphi_\mu \otimes \psi_\nu) \left(\sum_{i=1}^{n} g_i \otimes y_i^{\mu\nu} \right)| \le 3n^2 m^2 n\ell c\|P\|.$$

The two estimates together yield the one in the Lemma. ∎

8.5 LEMMA. *Let G and H be subspaces in Banach spaces X and Y respectively. Let α be a uniform reasonable crossnorm on $X \otimes Y$. Assume that X and Y possess biorthonormal systems $\{x_i, \varphi_i\}_1^k$ and $\{y_j, \psi_j\}_1^\ell$ respectively such that $\sum_{i=1}^{k} x_i \in G$ and $\sum_{j=1}^{\ell} y_j \in H$. If either*

(i) $\alpha^*[\sum_{ij} \epsilon_{ij}(\varphi_i \otimes \psi_j)] \le 1$ whenever $|\epsilon_{ij}| = 1$ or

(ii) $\alpha[\sum_{ij} \epsilon_{ij}(x_i \otimes y_j)] \le 1$ whenever $|\epsilon_{ij}| = 1$,

then every projection P of $X \otimes_\alpha Y$ onto $G \overline{\otimes} Y + X \overline{\otimes} H$ satisfies the inequality

$$\|P\| \ge 3 - 2(k\ell)^{-1} \sum_{ij} (\varphi_i \otimes \psi_j)(P(x_i \otimes y_j)).$$

PROOF. Put $z_{ij} = x_i \otimes y_j$. Then

$$\sum_i Pz_{ij} = P\left(\sum_i x_i \otimes y_j \right) = P\left[(\sum_i x_i) \otimes y_j \right] = \sum_i x_i \otimes y_j = \sum_i z_{ij}.$$

Similarly, $\sum_j Pz_{ij} = \sum_j z_{ij}$. Thus,

$$(\varphi_\mu \otimes \psi_\nu) \left(\sum_i Pz_{ij} \right) = \sum_i (\varphi_\mu \otimes \psi_\nu)(x_i \otimes y_j)$$

$$= \sum_i \varphi_\mu(x_i)\psi_\nu(y_j) = \sum_i \delta_{\mu i}\delta_{\nu j} = \delta_{\nu j}.$$

Similarly,

$$\sum_j (\varphi_\mu \otimes \psi_\nu)(Pz_{ij}) = \delta_{\mu i}.$$

It then follows that

$$\sum_{ij} (\varphi_\mu \otimes \psi_\nu)(Pz_{ij}) = 1.$$

By 8.2, we conclude that

$$\sum_{ij} |(\varphi_\mu \otimes \psi_\nu)(Pz_{ij})| \geq 3 - 2(\varphi_\mu \otimes \psi_\nu)(Pz_{\mu\nu}).$$

Now assume hypothesis (i). Then

$$\|P\| \geq \alpha(Pz_{ij}) \geq \left(\sum_{\mu\nu} \epsilon_{\mu\nu} \varphi_\mu \otimes \psi_\nu \right)(Pz_{ij}).$$

For appropriate signs $\epsilon_{\mu\nu}$ this yields

$$\|P\| \geq \sum_{\mu\nu} |(\varphi_\mu \otimes \psi_\nu)(Pz_{ij})|.$$

By summing this inequality over all i and j and using an inequality above, we obtain

$$k\ell\|P\| \geq \sum_{ij\mu\nu} |(\varphi_\mu \otimes \psi_\nu)(Pz_{ij})| \geq 3k\ell - 2\sum_{\mu\nu} (\varphi_\mu \otimes \psi_\nu)(Pz_{\mu\nu}).$$

This is the inequality to be proved.

If we assume hypothesis (ii) instead of (i) we have

$$\|P\| \geq \alpha \left(P \sum_{ij} \epsilon_{ij} z_{ij} \right) \geq (\varphi_\mu \otimes \psi_\nu) \left(P \sum_{ij} \epsilon_{ij} z_{ij} \right).$$

For appropriate ϵ_{ij} this yields

$$\|P\| \geq \sum_{ij} |(\varphi_\mu \otimes \psi_\nu)(Pz_{ij})|.$$

After summing over μ and ν we have

$$k\ell\|P\| \geq \sum_{ij\mu\nu} |(\varphi_\mu \otimes \psi_\nu)(Pz_{ij})|.$$

The remainder of the proof is as before. ∎

8.6 THEOREM. *Let G and H be finite-dimensional subspaces in Banach spaces X and Y respectively. Let α be a uniform reasonable crossnorm on $X \otimes Y$. Assume that corresponding to each natural number k there exist biorthonormal systems $\{x_i, \varphi_i\}_{i=1}^k$ for X and $\{y_i, \psi_i\}_{i=1}^k$ for Y such that $\sum_{i=1}^k x_i \in G$, $\sum_{i=1}^k y_i \in H$, and*

$$\sum_{ij} |(\varphi_i \otimes \psi_j)(z)| \leq \alpha(z) \qquad z \in X \otimes_\alpha Y.$$

Then each projection of $X \otimes_\alpha Y$ onto $G \otimes Y + X \otimes H$ has norm at least 3.

PROOF. Let $g \in G$ and put $w = g \otimes y_\nu$. Then

$$\sum_i |\varphi_i(g)| = \sum_{i,j} |\varphi_i(g)\psi_j(y_\nu)| = \sum_{i,j} |(\varphi_i \otimes \psi_j)(g \otimes y_\nu)| \le \alpha(g \otimes y_\nu) = \|g\|.$$

Similarly one proves that

$$\sum_j |\psi_j(h)| \le \|h\| \qquad (h \in H).$$

Hypothesis (i) of 8.5 is fulfilled because

$$\alpha^* \left[\sum_{ij} \epsilon_{ij}(\varphi_i \otimes \psi_j) \right] = \sup_{\alpha(z)=1} \left| \sum_{ij} \epsilon_{ij}(\varphi_i \otimes \psi_j)z \right|$$
$$\le \sup_{\alpha(z)=1} \sum_{ij} |(\varphi_i \otimes \psi_j)(z)| \le 1.$$

Now let P be a projection as described in the Theorem. By 8.5,

$$\|P\| \ge 3 - 2k^{-2} \sum_{ij} (\varphi_i \otimes \psi_j)(P(x_i \otimes y_j)).$$

By 8.4, the sum in this inequality is $\mathcal{O}(k)$ as $k \to \infty$. Hence in the limit we obtain $\|P\| \ge 3$. ∎

8.7 LEMMA. Let $\varphi_1, \dots, \varphi_n \in X^*$ and $\psi_1, \dots, \psi_m \in Y^*$, where X and Y are Banach spaces. Assume that

(i) $\sum_{\mu=1}^n |\varphi_\mu(x)| \le \|x\|$ $\qquad (x \in X)$

(ii) $\sum_{\nu=1}^m |\psi_\nu(y)| \le \|y\|$ $\qquad (y \in Y)$.

Then for all $z \in X \otimes_\gamma Y$ we have

$$\sum_{\mu=1}^n \sum_{\nu=1}^m |(\varphi_\mu \otimes \psi_\nu)(z)| \le \gamma(z).$$

PROOF. It suffices to prove the inequality for an arbitrary z in the uncompleted tensor product $X \otimes Y$. Let z be such an element, and let one of its representations be

$$z = \sum_{i=1}^k x_i \otimes y_i.$$

Then

$$\sum_{\mu\nu} |(\varphi_\mu \otimes \psi_\nu)(z)| = \sum_{\mu\nu} \left| \sum_i \varphi_\mu(x_i)\psi_\nu(y_i) \right|$$
$$\le \sum_{\mu\nu i} |\varphi_\mu(x_i)| \, |\psi_\nu(y_i)| = \sum_i \sum_\mu |\varphi_\mu(x_i)| \sum_\nu |\psi_\nu(y_i)|$$
$$\le \sum_i \|x_i\| \, \|y_i\|.$$

96

If we now take an infimum over all representations of the element z and apply the definition of the norm γ, the result is the inequality to be proved. ∎

8.8 THEOREM. *Let S and T be finite measure spaces without atoms. Let G and H be finite-dimensional subspaces containing the constants in $L_1(S)$ and $L_1(T)$ respectively. Then every projection of $L_1(S \times T)$ onto $L_1(S) \otimes H + G \otimes L_1(T)$ has norm at least 3.*

PROOF. By 10.13, there exists for each integer k a measurable partition of S into k sets of equal measure. Let $\{S_1, \ldots, S_k\}$ be such a partition. Define

$$x_i(s) = c_i(s)/\mu(S_i)$$

where c_i denotes the characteristic function of S_i. Define $\varphi_i \in L_1(S)^*$ by

$$\varphi_i(x) = \int x(s)c_i(s)\, ds.$$

One sees at once that $\{x_i, \varphi_i\}_{i=1}^k$ is a biorthonormal system in $L_1(S)$. Furthermore,

$$\sum_{i=1}^k x_i \in G$$

since G contains constants. It is also apparent that

$$\sum_i |\varphi_i(x)| = \sum_i \left| \int c_i(s)x(s)\, ds \right| \leq \sum_i \int c_i(s)|x(s)|\, ds = \|x\|.$$

In the same way, we construct a biorthonormal system $\{y_i, \psi_i\}$ for $L_1(T)$. By Lemma 8.7, and 1.16,

$$\sum_i \sum_j |(\varphi_i \otimes \psi_j)(z)| \leq \|z\| \qquad z \in L_1(S \times T).$$

By 8.6, the assertion of this theorem follows. ∎

8.9 COROLLARY. *Assume the hypotheses of the preceding theorem. If G and H are the ranges of norm-1 projections, then there exists a minimal projection of $L_1(S \times T)$ onto $L_1(S) \otimes H + G \otimes L_1(T)$, and it has norm 3.*

PROOF. Combine 8.1 and 8.8. ∎

8.10 DEFINITION. *Let G be a finite-dimensional subspace of a Banach space X. We say that the pair (G, X) has "Property B" if for each $\epsilon > 0$ there exists a biorthonormal system $\{x_i, \varphi_i\}_1^k$ such that*

    (i) For some $\varphi \in X^*$, $\|\varphi\| \leq 1$ and $\sum_{i=1}^k |(\varphi_i - \varphi)(g)| \leq k\epsilon \|g\|$ on G;

    (ii) $\sum_{i=1}^k x_i \in G$ and $\|\sum_{i=1}^k x_i\| \leq k\epsilon$.

8.11 THEOREM. Let (G, X) and (H, Y) have Property B, and let α be a uniform reasonable crossnorm on $X \otimes Y$. Assume that the biorthonormal systems referred to in the definition of Property B have the property

$$\alpha\left[\sum_{i=1}^{k}\sum_{j=1}^{\ell}\epsilon_{ij}(x_i \otimes y_j)\right] \le 1 \qquad |\epsilon_{ij}| = 1.$$

Then the subspace $W = G \otimes Y + X \otimes H$ satisfies $\lambda(W, X \otimes_\alpha Y) \ge 3$.

PROOF. Let $P : X \otimes_\alpha Y \twoheadrightarrow W$ be a projection. Let $\delta > 0$. We shall prove that $\|P\| > 3 - \delta$.

Select norm-1 bases $\{g_1, \ldots, g_n\}$ for G and $\{h_1, \ldots, h_m\}$ for H. Select $\epsilon > 0$ so small that

(1)
$$4\left\{\sum_{i=1}^{n}\|\mathcal{G}_i\| + \sum_{j=1}^{m}\|\mathcal{H}_j\|\right\}\|P\|\epsilon < \delta$$

where $\mathcal{G}_i$ and $\mathcal{H}_j$ are the operators referred to in 11.6. Let $\{x_i, \varphi_i\}_1^k$ and $\{y_j, \psi_j\}_1^\ell$ be the biorthonormal systems that exist because of Property B.

Define $z_{ij} = x_i \otimes y_j$ and $w_{ij} = Pz_{ij}$ for $1 \le i \le k$ and $1 \le j \le \ell$. As in the proof of 8.5, by the crossnorm property of α, and by 8.10

$$\sum_{i=1}^{k} w_{ij} = \left(\sum_{i=1}^{k} x_i\right) \otimes y_j \qquad \alpha\left(\sum_{i=1}^{k} w_{ij}\right) = \left\|\sum_{i=1}^{k} x_i\right\| \le \epsilon k$$

$$\sum_{j=1}^{\ell} w_{ij} = x_i \otimes \sum_{j=1}^{\ell} y_j \qquad \alpha\left(\sum_{j=1}^{\ell} w_{ij}\right) = \left\|\sum_{j=1}^{\ell} y_j\right\| \le \epsilon\ell.$$

By Lemma 11.6,

(2)
$$w_{ij} = \sum_{\mu=1}^{n} g_\mu \otimes \mathcal{G}_\mu w_{ij} + \sum_{\nu=1}^{m} \mathcal{H}_\nu w_{ij} \otimes h_\nu.$$

By 8.5,

(3)
$$\|P\| \ge 3 - 2(k\ell)^{-1}\sum_{i=1}^{k}\sum_{j=1}^{\ell}(\varphi_i \otimes \psi_j)(w_{ij}).$$

In order to estimate the double summation appearing in (3), we split w_{ij} into the two parts given in (2). The analyses of the two parts are similar, and only one is presented.

Taking the functional φ given by Property B, we have

$$\sum_{i=1}^{k}\sum_{j=1}^{\ell}(\varphi_i \otimes \psi_j)\left[\sum_{\mu=1}^{n} g_\mu \otimes \mathcal{G}_\mu w_{ij}\right] = \sum_{ij\mu} \varphi_i(g_\mu)\psi_j(\mathcal{G}_\mu w_{ij})$$

$$= \sum_{ij\mu}\varphi(g_\mu)\psi_j(\mathcal{G}_\mu w_{ij}) + \sum_{ij\mu}(\varphi_i - \varphi)(g_\mu)\psi_j(\mathcal{G}_\mu w_{ij})$$

$$\leq \sum_{j\mu}\varphi(g_\mu)\psi_j\left(\mathcal{G}_\mu\left(\sum_i w_{ij}\right)\right) + \sum_{ij\mu}|(\varphi_i - \varphi)(g_\mu)|\,\|\psi_j\|\,\|\mathcal{G}_\mu\|\alpha(w_{ij})$$

$$\leq \sum_{j\mu}\|\varphi\|\,\|g_\mu\|\,\|\psi_j\|\,\|\mathcal{G}_\mu\|\alpha\left(\sum_i w_{ij}\right) + \sum_{ij\mu}|(\varphi_i - \varphi)(g_\mu)|\,\|\mathcal{G}_\mu\|\,\|P\|$$

$$\leq \sum_{j\mu}\|\mathcal{G}_\mu\|\epsilon k + \|P\|\sum_{j\mu}k\epsilon\|g_\mu\|\,\|\mathcal{G}_\mu\|$$

$$= \epsilon k\ell\sum_\mu\|\mathcal{G}_\mu\| + \epsilon k\ell\|P\|\sum_\mu\|\mathcal{G}_\mu\|$$

$$= \epsilon k\ell\sum_\mu\|\mathcal{G}_\mu\|(1 + \|P\|)$$

$$\leq 2\epsilon k\ell\|P\|\sum_\mu\|\mathcal{G}_\mu\|.$$

The analysis of the other term leads to an upper bound of $2\epsilon k\ell\|P\|\sum_\nu\|\mathcal{X}_\nu\|$. The sum of the two terms is bounded above by

$$2\epsilon k\ell\|P\|\left\{\sum_{\mu=1}^{n}\|\mathcal{G}_\mu\| + \sum_{\nu=1}^{m}\|\mathcal{X}_\nu\|\right\}.$$

Hence by (3) and (1),

$$\|P\| \geq 3 - 4\epsilon\|P\|\left\{\sum_{\mu=1}^{n}\|\mathcal{G}_\mu\| + \sum_{\nu=1}^{m}\|\mathcal{X}_\nu\|\right\} > 3 - \delta. \quad\blacksquare$$

8.12 LEMMA. *Let S be an infinite compact Hausdorff space. Let G be a finite-dimensional subspace containing the constants in $C(S)$. Then the pair $G, C(S)$ has Property B as defined in 8.10.*

PROOF. Let $\epsilon > 0$. We shall construct a biorthonormal system $\{x_i, \varphi_i\}_1^k$ with the characteristics required for Property B.

Since S is compact Hausdorff and infinite, it has an ω-accumulation point σ [**111**, p. 138]. Thus each neighborhood of σ contains infinitely many points of S. Define

$$U_i = \{s \in S : |g(\sigma) - g(s)| < 2^{-i} \quad \text{for all } g \text{ in the unit cell of } G\}.$$

Since the unit cell of G is equicontinuous, U_i is a neighborhood of σ. Let $k > 1/\epsilon$. Select distinct points $s_1, \ldots, s_k$ with $s_i \in U_i$. Let $V_1, \ldots, V_k$ be an open cover of S such that $s_j \notin V_i$ when $i \neq j$. Let $x_1, \ldots, x_k$ form a partition of unity subordinate to $\{V_i\}$ [148, p. 41]. Then $0 \leq x_i \leq 1$, $x_i(s_j) = \delta_{ij}$, $\sum_{i=1}^{k} x_i = 1$, and $\|x_i\| = 1$. Since

$$\sum_{i=1}^{k} |g(s_i) - g(\sigma)| < \sum_{i=1}^{k} 2^{-i} < 1 < k\epsilon \qquad (g \in G, \ \|g\| \leq 1)$$

all the required properties are possessed by $\{x_i, \hat{s}_i\}_1^k$. (Here $\hat{s}_i$ denotes the evaluation functional corresponding to the point s_i.) ∎

8.13 THEOREM. *Let S and T be infinite compact Hausdorff spaces, and let G and H be finite-dimensional subspaces containing constants in $C(S)$ and $C(T)$ respectively. Then every projection of $C(S \times T)$ onto $G \otimes C(T) + C(S) \otimes H$ has norm at least 3.*

PROOF. By the preceding lemma, the pairs $G, C(S)$ and $H, C(T)$ have Property B. In the proof of that lemma, the functions x_i constructed in $C(S)$ and the functions y_i constructed in the same way in $C(T)$ have the property

$$\lambda \left[\sum_{ij} \epsilon_{ij}(x_i \otimes y_j) \right] \leq 1 \qquad |\epsilon_{ij}| = 1.$$

Hence 8.11 applies. ∎

8.14 COROLLARY. *Assume the hypotheses of the preceding theorem. If G and H are norm-1 complemented, then there is a minimal projection of norm 3 from $C(S \times T)$ onto $G \otimes C(T) + C(S) \otimes H$.*

8.15 EXAMPLE. Boris Shekhtman communicated to us an example of a subspace

$$W = G \otimes C(T) + C(S) \otimes H$$

having a norm-1 projection. Let $S = [0, a] \cup \{1\}$ and $T = [0, b] \cup \{1\}$, where $0 < a < 1$ and $0 < b < 1$. In $C(S)$, let G be the one-dimensional subspace generated by the function $g(s) = s$. In $C(T)$, let H be generated by $h(t) = t$. Two projections are defined by

$$P : C(S) \to G \qquad Px = x(1)g$$
$$Q : C(T) \to H \qquad Qy = y(1)h.$$

Then let $\overline{P} = P \otimes_\lambda I$, $\overline{Q} = I \otimes_\lambda Q$, and $L = \overline{P} \oplus \overline{Q}$. The map L is a projection of $C(S \times T)$ onto W, and a computation of its norm is now necessary. First, if $z \in C(S \times T)$ and $\|z\| = 1$ then

$$\|Lz\| = \sup_{s,t} |z(1, t)g(s) + z(s, 1)h(t) - z(1, 1)h(t)g(s)|$$

$$= \sup_{s,t} |sz(1, t) + tz(s, 1) - stz(1, 1)|.$$

The set $S \times T$ is shown in the figure.

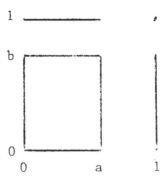

Thus,

$$\|Lz\| = \max\{\sup_{t \in T} |z(1,t) + tz(1,1) - tz(1,1)|, \ \sup_{s \in S} |sz(1,1) + z(s,1) - sz(1,1)|,$$

$$\sup_{(s,t) \in [0,a] \times [0,b]} |sz(1,t) + tz(s,1) - stz(1,1)|\}$$

$$= \max\{\sup_{t \in T} |z(1,t)|, \ \sup_{s \in S} |z(s,1)|, \ \sup_{(s,t) \in [0,a] \times [0,b]} |sz(1,t) + tz(s,1) - stz(1,1)|\}$$

$$\leq \max\{1, a+b+ab\}.$$

If $a = b = 1/3$, we get $\|L\| = 1$. ∎

8.16 LEMMA. *Let S be a measure space without atoms. Let G be a finite-dimensional subspace containing constants in $L_\infty(S)$. Then the pair $G, L_\infty(S)$ has Property B as defined in 8.10.*

PROOF. Since S has no atoms, there is an infinite sequence of measurable sets $S_1, S_2, \ldots$ in S such that $0 < \mu(S_i) < \infty$ and $S_i \cap S_j = \Box$ when $i \neq j$ (10.11). Define functionals φ_i on $L_\infty(S)$ by

$$\varphi_i(x) = \mu(S_i)^{-1} \int_{S_i} x(s) \, ds.$$

Then $\|\varphi_i\| = \varphi_i(1) = 1$. By the weak*-compactness of the unit cell in $L_\infty(S)^*$ it follows that the sequence $\{\varphi_i\}_1^\infty$ has a weak*-cluster point, φ, and $\|\varphi\| \leq 1$. Since G is finite dimensional, the sets

$$U_i = \{\psi \in L_\infty(S)^* : |\varphi(g) - \psi(g)| < 2^{-i} \quad \text{for } g \in G, \ \|g\| \leq 1\}$$

are weak*-neighborhoods of φ. Select an increasing sequence of integers $\nu_1, \nu_2, \ldots$ such that $\varphi_{\nu_i} \in U_i$.

Now let $\epsilon > 0$. Select $k > 1/\epsilon$. Define $x_2, \ldots, x_k$ in $L_\infty(S)$ as the characteristic functions of $S_{\nu_2}, \ldots, S_{\nu_k}$. Let $A = S \backslash (S_{\nu_2} \cup \cdots \cup S_{\nu_k})$, and let x_1 be the characteristic function of A. Then $\{x_i, \varphi_{\nu_i}\}_1^k$ is a biorthonormal system. Furthermore, $\sum_{i=1}^k x_i = 1 \in G$. Finally, if $g \in G$ and $\|g\| \le 1$, then

$$\frac{1}{k}\sum_{i=1}^k |\varphi_{\nu_i}(g) - \varphi(g)| \le \frac{1}{k}\sum_{i=1}^k 2^{-i} < \frac{1}{k} < \epsilon. \quad\blacksquare$$

8.17 THEOREM. *Let S and T be σ-finite measure spaces without atoms. Let G and H be finite-dimensional subspaces containing constants in $L_\infty(S)$ and $L_\infty(T)$ respectively. Then each projection of $L_\infty(S) \otimes_\lambda L_\infty(T)$ onto the subspace $G \otimes L_\infty(T) + L_\infty(S) \otimes H$ has norm at least 3.*

PROOF. The pairs $G, L_\infty(S)$ and $H, L_\infty(T)$ have Property B by 8.16. The biorthonormal systems $\{x_i, \varphi_i\}_1^k$ and $\{y_i, \psi_i\}_1^\ell$ which one constructs as in the proof of that lemma have the additional property

$$\left\| \sum_{ij} \epsilon_{ij} x_i \otimes y_j \right\| \le 1 \qquad |\epsilon_{ij}| = 1.$$

By 1.53,

$$\lambda\left(\sum_{ij} \epsilon_{ij} x_i \otimes y_j \right) = \left\| \sum_{ij} \epsilon_{ij} x_i \otimes y_j \right\| \le 1.$$

By 8.11, the result follows. $\quad\blacksquare$

8.18 COROLLARY. *Assume the hypotheses of 8.17. Then each projection of $L_\infty(S \times T)$ onto*

$$G \otimes L_\infty(T) + L_\infty(S) \otimes H$$

has norm at least 3.

MINIMAL PROJECTIONS

In the previous chapter we discussed lower bounds for projection constants. In this chapter we show that some of those lower bounds are attained, and also investigate minimal projections in a greater variety of spaces. The key to much of what takes place is a general theorem of Rudin [**150**]. We begin by establishing this result.

The setting for this theorem is as follows. We have a Banach space X and a compact topological group G. Defined on X is a set $\mathcal{A}$ of bounded linear, bijective operators in such a way that $\mathcal{A}$ is algebraically isomorphic to G. The image of $g \in G$ under this isomorphism will be denoted by A_g. We shall assume that the map $(g, x) \mapsto A_g x$ from $G \times X$ into X is continuous. A subspace Y of X is said to be **invariant** under G if $A_g Y \subset Y$ for all $g \in G$. An operator $B : X \to X$ is said to **commute** with G if $B A_g = A_g B$ for all g in G. Throughout the remainder of the exposition we will write g in place of A_g and use such notation as $\|g\|$ when we mean $\|A_g\|$.

9.1 THEOREM. *Let X and G satisfy the above hypotheses, and let Y be a closed subspace of X which is invariant under G. If there exists a bounded projection P of X onto Y, then there exists a bounded projection Q of X onto Y which commutes with G.*

PROOF. We define the sets $E_{k,x} \subset G$ by

$$E_{k,x} = \{g \in G : \|gx\| \leq k \|x\|\}$$

where $x \in X$ and $k = 1, 2, 3, \ldots$. Since the mapping $g \mapsto gx$ is continuous $E_{k,x}$ is closed. Then the intersection of all such sets over $x \in X$, which we shall denote by E_k, is also closed. Clearly,

$$E_k = \{g \in G : \|g\| \leq k\},$$

and $G = \cup_{k=1}^{\infty} E_k$. The Baire theorem may now be applied to conclude that there exists an m and an open set $V \subset E_m$. Now each element of G is in some translate of V and so the translates gV form an open cover of G. Since G is compact, finitely many of these translates cover G, say $g_1 V, \ldots, g_n V$. Now for any $g \in G$ we have $g \in g_i V$ for some i; thus for some $v \in V$

$$\|g\| = \|g_i v\| \leq \max_{1 \leq j \leq n} \|g_j\| \|v\| \leq Mm.$$

Hence the operators in $\mathcal{A}$ are uniformly bounded. Define an operator Q on X by

$$Qx = \int_G g^{-1} Pgx \, dg.$$

Here dg denotes the Haar measure on G, normalised so that the measure of G is one.

Now for a fixed x, $g^{-1}Pgx$ is a continuous mapping from G into X. Hence Q is well-defined, linear, and $\|Q\| \leq Mm\,\|P\|$.

For any $x \in X$ we have $Pgx \in Y$ for all $g \in G$ and, since Y is invariant under G, $g^{-1}Pgx \in Y$. Since Y is closed, it follows that $Qx \in Y$. Furthermore, if $x \in Y$, then $gx \in Y$ and so $Pgx = gx$ and $g^{-1}Pgx = x$. Hence $Qx = x$, and we have established that Q is a projection of X onto Y.

Finally, we show that Q commutes with G. Fix $g_0 \in G$ and put $h = gg_0$ so that $g^{-1} = g_0 h^{-1}$. Using the fact that Haar measure is translation-invariant we have

$$Qg_0 x = \int_G g^{-1} Pgg_0 x \, dg = \int g_0 h^{-1} Ph \, dh = g_0 Qx. \quad \blacksquare$$

We shall now take S and T to be finite sets which we shall identify as

$$S = \{1, 2, \ldots, n\} \quad \text{and} \quad T = \{1, 2, \ldots, m\}.$$

Initially, we consider the linear space X of all real-valued functions on $S \times T$. Let U be the subspace of X consisting of functions which depend only on the s-variable while V is the subspace of functions depending on the t-variable. Let $W = U + V$. Of importance in the following discussion will be the fact that a function $x \in X$ is in W if and only if x satisfies the "four-point rule":

(1) $$x(i,j) = x(k,j) + x(i,\ell) - x(k,\ell) \quad 1 \leq i, k \leq n, \quad 1 \leq j, \ell \leq m.$$

This is established as follows. If $x \in U$ or $x \in V$, it is elementary to verify (1). Hence (1) holds for $x \in W$. On the other hand, if x belongs to X and satisfies (1), and if (k, ℓ) is a fixed point in $S \times T$, then (1) displays x as the sum of a function of j, viz. $x(k, j)$, and a function of i, viz. $x(i, \ell) - x(k, \ell)$.

It will be convenient to use matrix terminology for the function x, so that, for example, the values $x(i, j), 1 \leq j \leq m$, will be referred to as the ith row of x. Define permutations π_{ij} which interchanges rows i and j in x and τ_{ij} which interchanges columns i and j.

These generate a finite group which we will denote by G. The elements of G are associated with transformations on X in the obvious way so that we will take A_π to be defined by

$$(A_\pi x)(i, j) = (x \circ \pi)(i, j) \quad (\pi \in G).$$

104

We will consider X to have a norm which constrains the maps A to be isometries. The hypotheses of 9.1 are now met. The subspace W is invariant under the transformations A_π. The mapping $(x, g) \mapsto A_g x$ is continuous if G is given the discrete topology, since for any $x_0 \in X$ and $g \in G$ the set

$$\{(x, g) : \|x_0 - x\| < \epsilon\}$$

is carried by this mapping into the set

$$\{z : \|A_g x_0 - z\| < \epsilon\}.$$

In fact we claim a stronger result than 9.1 holds.

9.2 LEMMA. *There is a unique projection $Q : X \to W$ which commutes with G.*

PROOF. Take e_{rs} to be the function defined by

$$e_{rs}(i, j) = \delta_{ri} \delta_{sj}$$

and set

$$a_{rs} = Q e_{rs} = a_{rs} = \sum_{i=1}^{n} \sum_{j=1}^{m} a_{rs}(i, j) e_{ij}.$$

Now $\pi_{rs} e_{11} = e_{11}$ whenever $r, s \geq 2$, and recalling that we require

$$\pi_{rs} a_{11} = \pi_{rs} Q e_{11} = Q \pi_{rs} e_{11} = Q e_{11} = a_{11}$$

we must have

$$a_{11}(r, j) = a_{11}(s, j), \quad 2 \leq r, s \leq n, \quad 1 \leq j \leq m.$$

By a similar argument using τ_{rs} we obtain

$$a_{11}(i, r) = a_{11}(i, s), \quad 1 \leq i \leq n, \quad 2 \leq r, s \leq m.$$

These two equations give

$$a_{11}(i, j) = \begin{cases} b & i = j = 1 \\ c & 2 \leq i \leq n, \ j = 1 \\ d & i = 1, \quad 2 \leq j \leq m \\ e & 2 \leq i \leq n, \ 2 \leq j \leq m \end{cases}.$$

Now using

$$a_{rs} = Q e_{rs} = Q \tau_{s1} \pi_{r1} e_{11} = \tau_{s1} \pi_{r1} Q e_{11} = \tau_{s1} \pi_{r1} a_{11}$$

105

we have

$$a_{rs}(i,j) = \begin{cases} b & i = r, \quad j = s \\ c & i \neq r, \quad j = s \\ d & i = r, \quad j \neq s \\ e & i \neq r, \quad j \neq s \end{cases}.$$

Now since $a_{rs} \in W$, the four-point rule gives

$$b + e = c + d.$$

Also the two equations

$$Q\left(\sum_{i=1}^{n} e_{i1}\right) = \sum_{i=1}^{n} e_{i1} \quad \text{and} \quad Q\left(\sum_{j=1}^{m} e_{ij}\right) = \sum_{j=1}^{m} e_{ij}$$

give

$$b + (n-1)c = 1$$
$$d + (n-1)e = 0,$$

and

$$b + (m-1)d = 1$$
$$c + (m-1)e = 0.$$

These five equations have the unique solution

$$b = \frac{n+m-1}{nm}, \quad c = \frac{m-1}{nm}, \quad d = \frac{n-1}{nm}, \quad e = \frac{-1}{nm},$$

which defines the projection Q which commutes with G. ∎

9.3 THEOREM. *If P is any projection from X onto W, then the projection R defined by*

$$Rx = \int_G g^{-1} Pgx \, dg$$

is a minimal projection of X onto W. (Recall that $\|g\| = 1$ for all g.)

PROOF. From the proof of 9.1 we know that R is a projection which commutes with G. From 9.2 we know that such a projection is unique. Hence using Q from 9.2, we have

$$Qx = \int_G g^{-1} Pgx \, dg,$$

for any projection P. Since the transformations associated with G have unit norm,

$$\|Qx\| \leq \int_G \|g^{-1} Pgx\| \, dx \leq \int_G \|g^{-1}\| \, \|P\| \, \|g\| \, \|x\| \, dx$$
$$\leq \|P\| \, \|x\|.$$

106

Taking a supremum over all x of norm 1, we obtain $\|Q\| \leq \|P\|$. ∎

9.4 COROLLARY. Let $S = \{1, \ldots, n\}$ and $T = \{1, \ldots, m\}$ with measures μ and ν giving $\mu(i) = 1/n$, and $\nu(j) = 1/m$. Then a minimal projection Q from $L_p(S \times T)$ onto $L_p(S) + L_p(T)$ is given by

$$(Qe_{rs})(i, j) = \begin{cases} (n + m - 1)/(nm) & i = r, \quad j = s \\ (m - 1)/(nm) & i \neq r, \quad j = s \\ (n - 1)/(nm) & i = r, \quad j \neq s \\ (-1)/(nm) & i \neq r, \quad j \neq s \end{cases}$$

where $e_{rs}(i, j) = \delta_{ri} \delta_{sj}$ $(1 \leq p \leq \infty)$.

PROOF. Observe first that permutations of the rows or columns of a function

$$x \in L_p(S \times T)$$

do not alter its norm. Hence the operators A_π have unit norm. Thus the projection Q calculated in the proof of 9.2 is a minimal one by 9.3. ∎

This result may be extended to the case where S and T are finite measure spaces. To do this we must first establish some notation and preliminary results. We shall assume from now on that (S, Σ, μ) and (T, Θ, ν) are finite, non-atomic measure spaces with $\mu(S) = \nu(T) = 1$. The product measure space $(S \times T, \Phi, \sigma)$ is then constructed in the usual way. We shall from now on use X to denote $L_p(S \times T)$ and W to denote $L_p(S) + L_p(T)$. The spaces in the previous case (9.4), when S and T are purely atomic and are each assumed to contain n points, will now be referred to as X_n and W_n. Our candidate for a minimal projection from X onto W will be

$$(P_0 x)(s, t) = \int_S x(a, t)\, d\mu(a) + \int_T x(s, b)\, d\nu(b) - \iint_{S \times T} x(a, b)\, d\sigma(a, b).$$

It is easy to see that if $p = 1$ or ∞ then $\|P_0\| \leq 3$. Then 8.8 and 8.18 give:

9.5 THEOREM. The projection P_0 defined above is a minimal projection from $L_p(S \times T)$ onto $L_p(S) + L_p(T)$ where $p = 1$ or ∞ and S and T are non-atomic measure spaces each having measure 1.

In the case of $L_\infty(S \times T)$, the hypothesis that S and T are finite in 9.5 can be weakened to σ-finite. Then a minimal projection is given by a formula like the one above, except that the integrals are over any sets $A, B, A \times B$, all of measure 1.

107

Now for a fixed value of n take $\{S_i\}_1^n$, $\{T_j\}_1^n$ to be measurable partitions of S and T respectively satisfying

$$\mu(S_i) = \nu(T_j) = 1/n, \quad 1 \le i, j \le n.$$

This is possible by virtue of 10.13. With the aid of these partitions we define operators $U_n : X \to X_n$ and $V_n : X_n \to X$ by

$$(U_n x)(i,j) = n^2 \int\!\!\int_{S_i \times T_j} x \, d\sigma \qquad (1 \le i, j \le n)$$

and

$$V_n z = \sum_{i,j=1}^{n} z(i,j) \chi_{S_i \times T_j}.$$

9.6 LEMMA. *The linear operators U_n, V_n defined above are both of unit norm for $1 \le p < \infty$.*

PROOF. Firstly,

$$\|V_n z\|^p = \int\!\!\int |V_n z|^p = \int\!\!\int \left| \sum_{i,j} z(i,j) \chi_{S_i \times T_j} \right|^p = \int\!\!\int \sum_{i,j} |z(i,j)|^p \chi_{S_i \times T_j}$$

$$= \sum_{ij} |z(i,j)|^p \int\!\!\int \chi_{S_i \times T_j} = \sum_{ij} \frac{1}{n^2} |z(i,j)|^p = \|z\|^p.$$

Secondly, we have

$$\|U_n x\|^p = \frac{1}{n^2} \sum_{i,j} |(U_n x)(i,j)|^p$$

$$= \frac{1}{n^2} \sum_{i,j} \left| n^2 \int\!\!\int_{S_i \times T_j} x \, d\sigma \right|^p$$

$$\le \frac{1}{n^2} \sum_{i,j} \left[\int\!\!\int_{S \times T} |x| n^2 \chi^2_{S_i \times T_j} \, d\sigma \right]^p.$$

Now an application of Hölder's inequality gives

$$\|U_n x\|^p \le \frac{1}{n^2} \sum_{i,j} \left[\int\!\!\int_{S \times T} |x|^p \chi^p_{S_i \times T_j} \, d\sigma \right] \left[\int\!\!\int_{S \times T} n^{2q} \chi^q_{S_i \times T_j} \, d\sigma \right]^{p/q}$$

where, $p^{-1} + q^{-1} = 1$. Now

$$\int\!\!\int_{S \times T} n^{2q} \chi^q_{S_i \times T_j} \, d\sigma = n^{2q-2}$$

and so

$$\|U_n x\|^p \le \frac{1}{n^2} \|x\|^p \, n^{(2q-2)pq^{-1}} = \|x\|^p.$$

Hence $\|U_n\| \le 1$ and it is easy to see, using the function x which takes the value 1 everywhere, that $\|U_n\| = 1$. $\blacksquare$

9.7 LEMMA. If P is any projection from X onto W then $U_n P V_n$ is a projection from X_n onto W_n.

PROOF. Suppose $z \in X_n$. Then since P has range W, we have $P V_n z \in W$. Hence $P V_n z = u + v$ for some $u \in L_p(S)$, $v \in L_p(T)$. Now

$$(U_n P V_n z)(i,j) + (U_n P V_n z)(\ell, k) = n^2 \int\!\!\int_{S_i \times T_j} P V_n z \, d\sigma + n^2 \int\!\!\int_{S_\ell \times T_k} P V_n z \, d\sigma$$

$$= n^2 \int\!\!\int_{S_i \times T_j} (u+v) \, d\sigma + n^2 \int\!\!\int_{S_\ell \times T_k} (u+v) \, d\sigma$$

$$= n \left\{ \int_{S_i} u \, d\mu + \int_{T_j} v \, d\nu + \int_{S_\ell} u \, d\mu + \int_{T_k} v \, d\nu \right\}$$

$$= n^2 \left\{ \int\!\!\int_{S_i \times T_k} u \, d\sigma + \int\!\!\int_{S_\ell \times T_j} v \, d\sigma + \int\!\!\int_{S_\ell \times T_j} u \, d\sigma + \int\!\!\int_{S_i \times T_k} v \, d\sigma \right\}$$

$$= (U_n P V_n z)(i,k) + (U_n P V_n z)(\ell, j).$$

Thus $U_n P V_n z$ satisfies the "four-point rule" whenever $z \in X_n$ and so the range of $U_n P V_n$ is contained in W_n. To see that $U_n P V_n$ projects onto W_n, take $z \in W_n$ where

$$z(i,j) = k_i, \qquad 1 \le i, j \le n.$$

Then

$$V_n z = \sum_{i,j=1}^n z(i,j) \chi_{S_i \times T_j} = \sum_{i,j=1}^n k_i \chi_{S_i \times T_j} = \sum_{i=1}^n k_i \chi_{S_i \times T}.$$

Thus $V_n z \in L_p(S)$ and so $P V_n z = V_n z$. Now

$$(U_n P V_n z)(i,j) = (U_n V_n z)(i,j) = n^2 \int\!\!\int_{S_i \times T_j} \sum_{i,j=1}^n z(i,j) \chi_{S_i \times T_j} \, d\sigma = k_i, \quad 1 \le i,j \le n.$$

Hence $U_n P V_n z = z$. A similar argument shows that if $z \in W_n$ where

$$z(i,j) = k_j, \qquad 1 \le i, j \le n,$$

then $U_n P V_n z = z$ and so $U_n P V_n$ acts as the identity on W_n. $\blacksquare$

109

9.8 LEMMA. Let Q and P_0 be defined as in 9.4 and 9.5. Then $Q = U_n P_0 V_n$.

PROOF. It will suffice to establish $Q e_{rk} = U_n P_0 V_n e_{rk}$ where $e_{rk}(i,j) = \delta_{ri}\,\delta_{kj}$. Observe that $V_n e_{rk} = \chi_{S_r \times T_k}$ and so

$$
P_0 V_n e_{rk}(s,t) = \int_S \chi_{S_r \times T_k}(a,t)\, d\mu + \int_T \chi_{S_r \times T_k}(s,b)\, d\nu - \iint_{S \times T} \chi_{S_r \times T_k}(a,b)\, d\sigma
$$

$$
= \begin{cases}
\dfrac{2}{n} - \dfrac{1}{n^2} & s \in S_r, \quad t \in T_k \\[2mm]
\dfrac{1}{n} - \dfrac{1}{n^2} & s \in S_r, \quad t \notin T_k \\[2mm]
\dfrac{1}{n} - \dfrac{1}{n^2} & s \notin S_r, \quad t \in T_k \\[2mm]
\dfrac{-1}{n^2} & s \notin S_r, \quad t \notin T_k
\end{cases}
$$

Hence

$$
(U_n P_0 V_n e_{rk})(i,j) = \begin{cases}
\dfrac{2n-1}{n^2} & i = r, \quad j = k \\[2mm]
\dfrac{n-1}{n^2} & i \neq r, \; j = k \quad \text{or} \quad i = r, \; j \neq k \\[2mm]
\dfrac{-1}{n^2} & i \neq r, \quad j \neq k
\end{cases}
$$

$$
= (Q e_{rk})(i,j). \quad \blacksquare
$$

9.9 THEOREM. Let S and T be nonatomic measure spaces each of measure 1. Then the projection

$$
P_0 : L_p(S \times T) \to L_p(S) + L_p(T)
$$

given by

$$
(P_0 x)(s,t) = \int_S x(a,t)\, d\mu(a) + \int_T x(s,b)\, d\nu(b) - \iint_{S \times T} x(a,b)\, d\mu(a)\, d\nu(b)
$$

is a minimal projection for $1 \leq p \leq \infty$.

PROOF. We shall only consider $1 < p < \infty$ since $p = 1, \infty$ have been covered in 9.5. Given $\epsilon > 0$ we choose $x \in X$ with $\|x\| = 1$ and

$$
\|P_0 x\| \geq \|P_0\| - \epsilon.
$$

By 10.15 we can find partitions $\{S_i\}_1^n$, $\{T_j\}_1^n$ of S and T such that

$$
\mu(S_i) = \nu(T_j) = \frac{1}{n}, \qquad 1 \leq i, j \leq n
$$

and such that there exists a function y of the form

$$y = \sum_{i,j=1}^{n} a_{ij} \, \chi_{S_i \times T_j} \quad \text{with } \|x - y\| < \epsilon.$$

Combining 9.7, 9.8 and 9.3, we have for any projection P from X onto W and for all $z \in X_n$

$$U_n P_0 V_n z = \int_G g^{-1} U_n P V_n gz \, dg.$$

Now $U_n y \in X_n$, and so

$$U_n P_0 V_n U_n y = \int_G g^{-1} U_n P V_n g U_n y \, dg.$$

Because of the form of y we have $V_n U_n y = y$, and so

$$U_n P_0 y = \int_G g^{-1} U_n P V_n g U_n y \, dg.$$

Now for $(s,t) \in S_r \times T_k$ we have

$$(P_0 y)(s,t) = \int_S \sum_{i,j} a_{ij} \, \chi_{S_i \times T_j}(a,t) \, d\mu + \int_T \sum_{i,j} a_{ij} \, \chi_{S_i \times T_j}(s,b) \, d\nu$$

$$- \int\!\!\int_{S \times T} \sum_{i,j} a_{ij} \, \chi_{S_i \times T_j}(a,b) \, d\sigma$$

$$= \int_S \sum_i a_{ik} \, \chi_{S_i \times T_k}(a,t) \, d\mu + \int_T \sum_j a_{rj} \, \chi_{S_r \times T_j}(s,b) \, d\nu - \frac{1}{n^2} \sum_{i,j} a_{ij}$$

$$= \frac{1}{n} \left(\sum_i a_{ik} + \sum_j a_{rj} - \frac{1}{n} \sum_{i,j} a_{ij} \right).$$

Hence $P_0 y = \sum_{i,j=1}^{n} b_{ij} \, \chi_{S_i \times T_j}$. We have

$$V_n U_n P_0 y = V_n \int_G g^{-1} U_n P V_n g U_n y \, dg$$

or

$$P_0 y = V_n \int_G g^{-1} U_n P V_n g U_n y \, dg.$$

Now

$$\|P_0\| \leq \|P_0 x\| + \epsilon \leq \|P_0 y\| + \|P_0(x-y)\| + \epsilon$$

$$\leq \|P_0 y\| + \|P_0\| \, \|x - y\| + \epsilon$$

$$\leq \|P_0 y\| + \epsilon(\|P_0\| + 1)$$

$$= \left\| V_n \int_G g^{-1} U_n P V_n g U_n y \, dg \right\| + \epsilon(\|P_0\| + 1)$$

$$\leq \|V_n\| \int_G \|g^{-1}\| \, \|U_n\| \, \|P\| \, \|V_n\| \, \|g\| \, \|U_n\| \, \|y\| \, dg + \epsilon(\|P_0\| + 1).$$

By 9.6, $\|U_n\| = \|V_n\| = 1$. From the proof of 9.4, $\|g\| = 1$. Hence

$$\|P_0\| \leq \|P\| \, \|y\| + \epsilon(\|P_0\| + 1)$$
$$\leq \|P\| \, (1 + \epsilon) + \epsilon(\|P_0\| + 1).$$

Since ϵ was arbitrary, we conclude that P_0 is indeed a minimal projection. ∎

CHAPTER 10

APPENDIX ON THE BOCHNER INTEGRAL

In the course of the previous chapters, we have frequently discussed approximation-theoretic questions involving measures and integrals. In particular, we have employed integrals of functions whose values lie in Banach spaces. These integrals, known as **Bochner** integrals, are the subject of this expository chapter. Our intention is only to provide an introduction to their basic theory. At the end of the chapter, there are a few results concerning classical measure theory which were used without proof in earlier chapters.

The Bochner integral has as essential constituents a measure space $(S, \mathcal{A}, \mu)$ and a Banach space X. Thus S is a set, $\mathcal{A}$ is a σ-algebra of subsets of S, and μ is a (countably additive nonnegative) measure on $\mathcal{A}$. As is customary, we refer to S as the measure space, since $\mathcal{A}$ and μ remain fixed in the background. The elements of $\mathcal{A}$ are the **measurable sets**. A measurable set of measure zero is termed a **null set**.

A function $f : S \to X$ is said to be **simple** if its range contains only finitely many points $x_1, x_2, \ldots, x_n$ in X, and if $f^{-1}(x_i)$ is measurable for $i = 1, 2, \ldots, n$. Such an f can be written

$$f = \sum_{i=1}^{n} x_i \, \chi_{E_i}$$

where χ_{E_i} is the characteristic function of the set $E_i = f^{-1}(x_i)$. A function $f : S \to X$ is said to be **strongly measurable** if there exists a sequence $\{f_n\}$ of simple functions with

$$\lim_{n \to \infty} \|f_n(s) - f(s)\| = 0$$

almost everywhere (*i.e.*, except on a null set.) The following few lemmas help to explain the concept of strong measurability.

10.1 LEMMA. *Let S be a measure space, X a Banach space, and f a strongly measurable function from S to X. For any open or closed set Y in X, $f^{-1}(Y)$ can be expressed as the union of a measurable set with a subset of a null set.*

PROOF. Let $\{f_n\}$ be a sequence of simple functions such that $f_n(s) \to f(s)$ on the set $S' = S \setminus N$, where N is a suitable null set. Let Y be closed in X, and define

$$E_{nk} = \{s \in S' : \operatorname{dist}(f_n(s), Y) \leq 1/k\}.$$

Each set E_{nk} is measurable, and one verifies easily that

$$f^{-1}(Y) \cap S' = \bigcap_{k=1}^{\infty} \bigcup_{m=1}^{\infty} \bigcap_{n=m}^{\infty} E_{nk}.$$

Hence $f^{-1}(Y) \cap S'$ is measurable.

If Y is an open set in X, then write

$$f^{-1}(Y) \cap S' = S' \setminus f^{-1}(X \setminus Y).$$

The set on the right side of this equation is measurable, by the first half of the proof. In either case, the proof is completed by writing

$$f^{-1}(Y) = [f^{-1}(Y) \cap S'] \cup [f^{-1}(Y) \cap N]. \quad \blacksquare$$

10.2 LEMMA. *Let S be a complete measure space, X a Banach space, and f a strongly measurable function from S into X. Then f is measurable in the classical sense.*

PROOF. We have to show that $f^{-1}(\mathcal{O})$ is measurable for every open set $\mathcal{O} \subset X$. From the previous lemma we can write $f^{-1}(\mathcal{O}) = A \cup B$, where B is measurable and A is contained in a null set. Since the measure is complete, A is measurable, and hence $f^{-1}(\mathcal{O})$ is also measurable. $\quad \blacksquare$

10.3 LEMMA. *If $f : S \to X$ is measurable in the classical sense and has essentially separable range, then f is strongly measurable.*

PROOF. Let N be a null set such that $f(S \setminus N)$ is separable. Let $\{x_n\}$ be a countable dense set in $f(S \setminus N)$. For each n, a simple function f_n is defined as follows. Put $X_n = \{x_1, \ldots, x_n\}$ and fix $s \in S \setminus N$. If $\operatorname{dist}(f(s), X_n) > 1$ put $f_n(s) = 0$. If $\operatorname{dist}(f(s), X_n) \leq 1$ let i be the largest integer in $\{1, \ldots, n\}$ for which $\operatorname{dist}(f(s), X_n) < 1/i$. Let j be the first integer for which $\|f(s) - x_j\| < 1/i$ and set $f_n(s) = x_j$.

In order to show that $f_n(s) \to f(s)$ for almost all $s \in S$, let s be any point in $S \setminus N$ and let $0 < \epsilon \leq 1$. Select an integer k such that $1/k < \epsilon$, and select $m > k$ so that $\|x_m - f(s)\| < 1/k$. If $n \geq m$ then $x_m \in X_n$ and $\operatorname{dist}(f(s), X_n) < 1/k$. Hence

$$\|f_n(s) - f(s)\| \leq 1/k < \epsilon.$$

Thus f is the limit almost everywhere of a sequence of simple functions, and consequently is strongly measurable. $\quad \blacksquare$

114

10.4 LEMMA. *If a sequence of strongly measurable functions from a finite measure space to a Banach space converges almost everywhere to a strongly measurable function, then for each $\epsilon > 0$ there is a measurable set of measure less than ϵ on the complement of which the convergence is uniform.*

PROOF. Let f and f_n be strongly measurable from S to X, with $f_n(s) \to f(s)$ almost everywhere. Let $\epsilon > 0$. We will construct a measurable set E, having measure at most ϵ, such that $f_n(s) \to f(s)$ uniformly on $S \setminus E$.

By removing a null set from S we obtain a set S' such that $f_n(s) \to f(s)$ everywhere on S'. The proof of 10.1 shows that by removing another null set we obtain a set S'' with the property that $S'' \cap (f_n - f)^{-1}(\mathcal{O})$ is measurable for each open set $\mathcal{O}$ in X and for each n.

As a result of these considerations, we conclude that the following sets are measurable:

$$E_n^m = \bigcap_{i=n}^{\infty} \{s \in S'' \ : \ \|f_i(s) - f(s)\| < 1/m\}.$$

Clearly $E_1^m \subset E_2^m \subset \cdots$. Since $f_n(s) \to f(s)$ on S'', we have $S'' = \cup_{n=1}^{\infty} E_n^m$ for each m. Since S'' is of finite measure, there exists, for each m, an integer $k(m)$ such that

$$\mu(S'' \setminus E_{k(m)}^m) < \epsilon/2^m.$$

Define the set

$$E = \bigcup_{m=1}^{\infty} (S'' \setminus E_{k(m)}^m).$$

Then E is measurable and has measure less than ϵ. If $s \in S'' \setminus E$ then $s \in E_{k(m)}^m$ for all m, and consequently the inequality

$$\|f_i(s) - f(s)\| < 1/m$$

is true for all $i \geq k(m)$. This establishes the uniform convergence

$$f_n(s) \to f(s) \quad \text{on} \quad S'' \setminus E. \quad \blacksquare$$

10.5 LEMMA. *Let S be a finite measure space, X a Banach space, and $f : S \to X$ a strongly measurable function. Then there exists a null set E such that $f(S \setminus E)$ is separable in X.*

115

PROOF. Since f is strongly measurable, there is a sequence of simple functions g_n such that $\|g_n(s) - f(s)\| \to 0$ for almost all $s \in S$. Using 10.4, we can find for each integer m a measurable set E_m such that $\mu(E_m) < 1/m$ and such that the sequence $\{g_n\}$ converges uniformly to f on $S \setminus E_m$. The range, G_n, of g_n is a finite set in X, and $\cup_{n=1}^{\infty} G_n$ is a countable set whose closure contains $f(S \setminus E_m)$. The latter set is therefore separable. Define $E = \cap_{m=1}^{\infty} E_m$. Then $\mu(E) = 0$. Also $f(S \setminus E)$ is separable, being equal to $\cup_{m=1}^{\infty} f(S \setminus E_m)$. ∎

10.6 THEOREM. *Let S be a finite measure space, X be a Banach space, and $f : S \to X$ a strongly measurable function. Then there exists a sequence of countably-valued strongly measurable functions which converges almost everywhere uniformly to f.*

PROOF. By removing a null set from S if necessary (see 10.5) we may assume that $f(S)$ is separable. Suppose $\{x_n\}_1^{\infty}$ is a countable dense subset of $f(S)$. For $s \in S$, define $n = n(m, s)$ to be the first integer such that $\|f(s) - x_n\| < \frac{1}{m}$. Put $f_m(s) = x_n$. Then for all $s \in S$, $\|f(s) - f_m(s)\| < \frac{1}{m}$. It remains to show that each f_m is strongly measurable. Fix m, and by 10.1 write

$$\{s \in S \ : \ \|f(s) - x_i\| < \tfrac{1}{m}\} = C_i \cup D_i$$

where D_i is measurable and C_i is contained in a null set, C_i' say. Now put $C = \cup_i C_i'$ and define, for each $k \geq 1$,

$$f_m^k(s) = \begin{cases} f_m(s) & \text{whenever } s \in S \setminus C \text{ and } f_m(s) \in \{x_1, \cdots, x_k\} \\ 0 & \text{otherwise.} \end{cases}$$

Then the range of f_m^k is $\{0, x_1, x_2, \ldots, x_k\}$, and, for $1 \leq i \leq k$,

$$\begin{aligned}
(f_m^k)^{-1}(x_i) &= \{s \in S \setminus C \ : \ f_m^k(s) = x_i\} \\
&= \{s \in S \setminus C \ : \ f_m(s) = x_i\} \\
&= \{s \in S \setminus C \ : \ \|f(s) - x_i\| < \tfrac{1}{m} \ \text{ and } \ \min_{1 \leq j < i} \|f(s) - x_j\| \geq \tfrac{1}{m}\} \\
&= (S \setminus C) \cap (C_i \cup D_i) \setminus \bigcup_{j<i}(C_j \cup D_j) \\
&= (S \setminus C) \cap D_i \setminus \bigcup_{j<i} D_j.
\end{aligned}$$

This last set is measurable. Also, $(f_m^k)^{-1}(0)$ is measurable, since

$$(f_m^k)^{-1}(0) = S \setminus \bigcup_{i=1}^{k}(f_m^k)^{-1}(x_i).$$

Thus each f_m^k is a simple function. It is clear that $f_m^k(s) \to f_m(s)$ for almost all $s \in S$ and so f_m is strongly measurable. ∎

Note that if f is a strongly measurable function having countable range $\{x_1, x_2, \ldots\}$, then for almost all s we have

$$f(s) = \sum_{i=1}^{\infty} c_i(s) x_i$$

where the c_i are characteristic functions of measurable sets. Indeed, by 10.1, we can write

$$f^{-1}(x_i) = [f^{-1}(x_i) \cap N] \cup [f^{-1}(x_i) \setminus N]$$

for a suitable null set N. Then c_i can be defined as the characteristic function of $f^{-1}(x_i) \setminus N$.

Now let $(S, \mathcal{A}, \mu)$ be a finite measure space, and f a simple function from S to a Banach space X. Write

$$f(s) = \sum_{i=1}^{n} c_i(s) x_i$$

where the c_i are characteristic functions of sets $E_i \in \mathcal{A}$. Then we define the integral of f over any measurable set E by

$$\int_E f(s) ds = \sum_{i=1}^{n} \mu(E \cap E_i) x_i.$$

This definition is independent of the representation of f.

10.7 LEMMA. *For any simple function $f : S \to X$,*

$$\left\| \int_E f(s) ds \right\| \le \int_E \| f(s) \| ds.$$

PROOF. It suffices to give the proof when $E = S$. Let f be as above, assuming in addition that the sets E_i are mutually disjoint. Then

$$\left\| \int f(s) ds \right\| = \left\| \sum_{i=1}^{n} \mu(E_i) x_i \right\| \le \sum_{i=1}^{n} \mu(E_i) \| x_i \|$$

$$= \int \sum_{i=1}^{n} c_i(s) \| x_i \| ds = \int \left\| \sum_{i=1}^{n} c_i(s) x_i \right\| ds$$

$$= \int \| f(s) \| ds. \quad \blacksquare$$

10.8 DEFINITION. *Let S be a finite measure space and X a Banach space. A strongly measurable function $f : S \to X$ is said to be Bochner integrable if there exists a sequence of simple functions f_n such that*

(i) $\lim_n \int \| f_n(s) - f(s) \| ds = 0$.

In this case we define

(ii) $\int_E f(s) ds = \lim_n \int_E f_n(s) ds$.

In order to see that the limit in (ii) exists, use 10.7 to write

$$\left\| \int_E f_n(s)\,ds - \int_E f_m(s)\,ds \right\| \le \int_S \|f_n(s) - f_m(s)\|\,ds$$

$$\le \int_S \|f_n(s) - f(s)\|\,ds + \int_S \|f(s) - f_m(s)\|\,ds.$$

Thus the sequence in (ii) is a Cauchy sequence in X.

Different sequences satisfying (i) lead to the same value for the integral in (ii) by a similar argument.

10.9 THEOREM. *Let S be a finite measure space and X a Banach space. A strongly measurable function $f : S \to X$ is Bochner integrable if and only if $\int_S \|f(s)\|\,ds$ is finite.*

PROOF. Assume that f is Bochner integrable, and let $\{f_n\}$ be a sequence of simple functions with $\int_S f(s)\,ds = \lim_n \int_S f_n(s)\,ds$. Then

$$\int_S \|f(s)\|\,ds \le \int_S \|f(s) - f_n(s)\|\,ds + \int_S \|f_n(s)\|\,ds$$

which is finite for all n.

Conversely, suppose f is strongly measurable, and $\int_S \|f(s)\|\,ds < \infty$. By 10.6 there exists a sequence $\{f_n\}$ of countably-valued strongly measurable functions and a null set E such that $\|f(s) - f_n(s)\| < 1/n$ for all $s \in S \setminus E$. Since

$$\|f_n(s)\| \le \|f(s)\| + 1/n$$

on $S \setminus E$, we have

$$\int_S \|f_n(s)\|\,ds \le \int_S \|f(s)\|\,ds + \mu(S)/n$$

and so $\int_S \|f_n(s)\|\,ds < \infty$. By the remark following 10.6, we can write the following equation, for all n and for almost all s:

$$f_n(s) = \sum_{m=1}^{\infty} x_{nm}\, c_{nm}(s),$$

where $x_{nm} \in X$ and c_{nm} is the characteristic function of a measurable set E_{nm}. Without loss of generality, we can assume

$$E_{ni} \cap E_{nj} = \square \quad \text{for} \quad i \ne j.$$

For each n chose p_n such that the set

$$F_n = \bigcup_{n=p_n+1}^{\infty} E_{nm}$$

has the property

$$\int_{F_n} \|f_n(s)\|\,ds < \mu(S)/n.$$

Define $g_n = \sum_{m=1}^{p_n} x_{nm} c_{nm}$. Then each g_n is a simple function with

$$\int\limits_S \|f(s) - g_n(s)\| \, ds \leq \int\limits_S \|f(s) - f_n(s)\| \, ds + \int\limits_S \|f_n(s) - g_n(s)\| \, ds \leq 2\mu(S)/n.$$

Hence f is Bochner integrable. ∎

There are many standard results from the scalar theory that can be carried over to the Bochner integral case. As an example we give the following theorem.

10.10 LEMMA. *Let S be a finite measure space, X a Banach space, and f a Bochner integrable function. Then*

(i) $\| \int f(s)ds \| \leq \int \|f(s)\| \, ds$

(ii) $\lim \int\limits_E f(s)ds = 0$ *as* $\mu(E) \to 0$, *uniformly in* E.

(iii) *If* $\int\limits_E f(s)ds = 0$ *for all measurable* E, *then* $f(s) = 0$ *a.e.*

PROOF. (i) This inequality is true for simple funcitons, by 10.7. Let $\{f_n\}$ be a sequence of simple functions such that

$$\lim \int \|f_n(s) - f(s)\| \, ds = 0.$$

From the inequalities

$$\left| \int \|f_n(s)\| \, ds - \int \|f(s)\| \, ds \right| \leq \int \left| \|f_n(s)\| - \|f(s)\| \right| ds$$

$$\leq \int \|f_n(s) - f(s)\|$$

we conclude that

$$\lim \int \|f_n(s)\| \, ds = \int \|f(s)\| \, ds.$$

Hence

$$\left\| \int f(s)ds \right\| = \left\| \lim \int f_n(s)ds \right\| = \lim \left\| \int f_n(s)ds \right\|$$

$$\leq \lim \int \|f_n(s)ds\| = \int \|f(s)\| \, ds.$$

(ii) Given $\epsilon > 0$, let g be a simple function such that

$$\int \|f(s) - g(s)\| \, ds < \epsilon/2.$$

Suppose that g has the representation

$$g = \sum_{i=1}^{n} x_i \, c_i$$

where each c_i is the characteristic function of a measurable set A_i. Let E be any measurable set such that

$$\mu(E) < \tfrac{1}{2}\epsilon / \sum_{i=1}^{n} \|x_i\|.$$

Then by (i),

$$\left\| \int_E f(s)ds \right\| \leq \int_E \|f(s)\| \, ds \leq \int_E \|g(s)\| \, ds + \epsilon/2$$

$$\leq \int_E \sum_{i=1}^n c_i(s)\|x_i\| \, ds + \epsilon/2$$

$$= \sum_{i=1}^n \|x_i\| \, \mu(A_i \cap E) + \epsilon/2$$

$$\leq \mu(E) \sum_{i=1}^n \|x_i\| + \epsilon/2 < \epsilon.$$

(iii)   Given $\epsilon > 0$, let g be a simple function such that

$$\int \|f(s) - g(s)\| \, ds < \epsilon.$$

Let

$$g(s) = \sum_{i=1}^n c_i(s)y_i$$

where c_i is the characteristic function of E_i and the sets E_i form a partition of S. Then

$$\int \|f(s)\| \, ds \leq \int \|g(s)\| \, ds + \epsilon$$

$$= \int \left\| \sum_{i=1}^n c_i(s)y_i \right\| \, ds + \epsilon$$

$$= \sum_{i1=}^n \mu(E_i)\|y_i\| + \epsilon$$

$$= \sum_{i=1}^n \left\| \int_{E_i} g(s)ds \right\| + \epsilon$$

$$= \sum_{i=1}^n \left\| \int_{E_i} (g(s) - f(s))ds + \int_{E_i} f(s)ds \right\| + \epsilon$$

$$= \sum_{i=1}^n \left\| \int_{E_i} (g(s) - f(s))ds \right\| + \epsilon$$

$$\leq \sum_{i=1}^n \int_{E_i} \|g(s) - f(s)\| \, ds + \epsilon$$

$$= \int \|g(s) - f(s)\| \, ds + \epsilon \leq 2\epsilon. \quad \blacksquare$$

We conclude this section about Bochner integrals by describing briefly some spaces of Bochner integrable functions. If $1 \leq p < \infty$ then $L_p(S, X)$ will denote the Banach space of (equivalence classes of) strongly measurable functions $f : S \to X$ such that

$$\int_S \|f(s)\|^p \, ds < \infty.$$

The norm in $L_p(S, X)$ is defined to be

$$\|f\|_p = \left\{ \int \|f(s)\|^p \, ds \right\}^{1/p}.$$

The essentially bounded strongly measurable functions $f : S \to X$ form the Banach space $L_\infty(S, X)$ with norm

$$\|f\|_\infty = \text{ess} \sup \|f(s)\|.$$

We now revert to the case $X = \mathbb{R}$ for the remainder of this appendix. Our purpose is to establish some measure theoretic results which were used in earlier chapters but were secondary to the main issues contained there.

Recall that an **atom** in a measure space S is a measurable set A of positive measure having the property that each of its measurable subsets has measure either 0 or $\mu(A)$. A measure space is said to be **non-atomic** if it contains no atoms.

10.11 LEMMA. Let S be a nontrivial non-atomic measure space. If $\epsilon > 0$, then there exists a measurable set E such that $0 < \mu(E) < \epsilon$.

PROOF. Since S is non-atomic, there exists a measurable set S' such that

$$0 < \mu(S') < \mu(S).$$

Then $\mu(S') < \infty$. Let E_1 be a measurable set in S' such that $0 < \mu(E_1) < \mu(S')$. Then $\mu(S' \setminus E_1) > 0$, and there exists a measurable set $E_2 \subset S' \setminus E_1$ such that

$$0 < \mu(E_2) < \mu(S' \setminus E_1).$$

Similarly, we find $E_3 \subset S' \setminus (E_1 \cup E_2)$ with

$$0 < \mu(E_3) < \mu(S' \setminus (E_1 \cup E_2)).$$

Proceeding in this way, we construct a disjoint sequence of sets $E_n \subset S'$ with $\mu(E_n) > 0$. Hence

$$\sum_{n=1}^{\infty} \mu(E_n) = \mu\left(\bigcup_{n=1}^{\infty} E_n \right) \leq \mu(S') < \infty.$$

Clearly $\mu(E_n) < \epsilon$ for sufficiently large n. ■

10.12 LEMMA. *Let S be a finite and non-atomic measure space. If $0 < \theta < \mu(S)$, then there is a measurable set E such that $\mu(E) = \theta$.*

PROOF. For any measurable set A, define the monotone set function

$$\beta(A) = \sup\{\mu(H) \; : \; H \text{ measurable}, \; H \subset A, \; \mu(H) \le \theta\}.$$

Using 10.11, select A_1 so that $0 < \mu(A_1) \le \theta$. Select $A_2 \subset S \setminus A_1$ so that

$$\tfrac{1}{2}\beta(S \setminus A_1) \le \mu(A_2) \le \theta.$$

Continuing inductively, we obtain a disjoint sequence of sets A_n such that

$$A_{n+1} \subset S \setminus \bigcup_{i=1}^{n} A_i$$

and

$$\tfrac{1}{2}\beta\left(S \setminus \bigcup_{i=1}^{n} A_i\right) \le \mu(A_{n+1}) \le \theta.$$

By the monotonicity of β,

$$\beta\left(S \setminus \bigcup_{i=1}^{\infty} A_i\right) \le \beta\left(S \setminus \bigcup_{i=1}^{n} A_i\right) \le 2\mu(A_{n+1}).$$

Since the A_i are disjoint, $\sum_{i=1}^{\infty} \mu(A_i) \le \mu(S)$, and hence $\mu(A_n) \to 0$. Therefore $\beta(S \setminus \cup_{i=1}^{\infty} A_i) = 0$. By 10.11,

$$\mu(S \setminus \cup A_i) = 0 \quad \text{and} \quad \sum_{i=1}^{\infty} \mu(A_i) = \mu(S) > \theta.$$

Hence there is a subsequence such that $\sum_{i=1}^{\infty} \mu(A_{n_i}) = \theta$. Consequently

$$\mu(\cup_{i=1}^{\infty} A_{n_i}) = \theta. \quad \blacksquare$$

10.13 LEMMA. *Let S be a finite non-atomic measure space. Then for each natural number n we can write $S = \cup_{i=1}^{n} S_i$ where $\mu(S_i) = \mu(S)/n$, and the S_i are pairwise disjoint.*

PROOF. The previous lemma shows that we can find S_1 in S with $\mu(S_1) = \mu(S)/n$. Apply this again to $S \setminus S_1$ to construct S_2, and so on. $\blacksquare$

10.14 LEMMA. *Let S be a finite, non-atomic measure space, and let $1 \le p < \infty$. If $x_1, \ldots, x_n \in L_p(S)$ and $\epsilon > 0$, then there exist simple functions $g_1, \ldots, g_n$ of the form*

$$g_i = \sum_{j=1}^{m} \lambda_{ij}\, c_j$$

where $\|x_i - g_i\|_p < \epsilon$, $c_\alpha c_\beta = 0$ if $\alpha \ne \beta$, $\sum_{j=1}^{m} c_j = 1$, and $\int c_j = \mu(S)/m$.

PROOF. Since the set of simple functions is dense in $L_p(S)$, there exist simple functions $h_1, \ldots, h_n$ such that $\|x_i - h_i\|_p < \epsilon/2$. By intersecting the measurable sets from which the h_i are constructed, we can assume that these functions have the form

$$h_i = \sum_{j=1}^{M} \alpha_{ij} u_j$$

where u_j is the characteristic function of a set E_j. The sets E_j can be assumed to form a partition of S. Choose N so that

$$N \epsilon^p \geq 2^p \max_i \left(\sum_{j=1}^{M} |\alpha_{ij}| \right)^p M \mu(S).$$

By 10.12, each E_j can be written as $E_j = A_j \cup B_j$, where $A_j \cap B_j = \square$, $\mu(B_j) < \mu(S)/N$, and $\mu(A_j)$ is a multiple of $\mu(S)/N$. By 10.12, each A_j can be written in the form

$$A_j = \bigcup_{k=1}^{r(j)} D_{jk}$$

where the sets $D_{j1}, D_{j2}, \ldots, D_{jr(j)}$ form a partition of A_j and have measure $\mu(S)/N$. Define $B = \cup_{j=1}^{M} B_j$. Since

$$B = S \setminus \bigcup_{j=1}^{M} A_j$$

we see that $\mu(B)$ is a multiple of $\mu(S)/N$. By 10.11, B can be written as

$$B = \bigcup_{k=1}^{r(0)} D_{0k}$$

where the sets D_{0k} form a partition of B and have measure $\mu(S)/N$. Thus the sets D_{jk} $(0 \leq j \leq M, 1 \leq k \leq r(j))$ partition S into disjoint sets of measure $\mu(S)/N$.

Now define simple functions

$$g_i = \sum_{j=0}^{M} \sum_{k=1}^{r(j)} \beta_{ijk} d_{jk}$$

where d_{jk} is the characteristic function of D_{jk}, and

$$\beta_{ijk} = \begin{cases} \alpha_{ij} & j \neq 0 \\ 0 & j = 0. \end{cases}$$

123

Now

$$\|h_i - g_i\|_p^p = \int \left| \sum_{\nu=1}^{M} \alpha_{i\nu}\, u_\nu - \sum_{j=0}^{M} \sum_{k=1}^{r(j)} \beta_{ijk}\, d_{jk} \right|^p$$

$$= \sum_{s=0}^{M} \sum_{t=1}^{r(s)} \int \left| \sum_{\nu=1}^{M} \alpha_{i\nu}\, u_\nu - \sum_{j=0}^{M} \sum_{k=1}^{r(j)} \beta_{ijk}\, d_{jk} \right|^p d_{st}$$

$$= \sum_{s=0}^{M} \sum_{t=1}^{r(s)} \int \left| \sum_{\nu=1}^{M} \alpha_{i\nu}\, u_\nu - \beta_{ist} \right|^p d_{st}$$

$$= \sum_{s=1}^{M} \sum_{t=1}^{r(s)} \int |\alpha_{is} - \beta_{ist}|^p d_{st} + \sum_{t=1}^{r(0)} \int \left| \sum_{\nu=1}^{M} \alpha_{i\nu}\, u_\nu \right|^p d_{st}$$

$$= \int_B \left| \sum_{\nu=1}^{M} \alpha_{i\nu}\, u_\nu \right|^p$$

$$\leq \int_B \left(\sum_{\nu=1}^{M} |\alpha_{i\nu}|\, |u_\nu| \right)^p$$

$$\leq \int_B \left(\sum_{\nu=1}^{M} |\alpha_{i\nu}| \right)^p$$

$$= \mu(B) \left(\sum_{\nu=1}^{M} |\alpha_{i\nu}| \right)^p$$

$$= \{M\mu(S)/N\} \left(\sum_{\nu=1}^{M} |\alpha_{i\nu}| \right)^p \leq \epsilon^p/2^p.$$

It follows that

$$\|x_i - g_i\|_p \leq \|x_i - h_i\|_p + \|h_i - g_i\|_p \leq \epsilon. \quad \blacksquare$$

10.15 LEMMA. Let $z \in L_p(S \times T)$, where $1 \leq p < \infty$ and S and T are finite non-atomic measure spaces. For each $\epsilon > 0$ there is a simple function f such that $\|z - f\|_p < \epsilon$ and such that f is of the form

$$f = \sum_{j=1}^{m} \sum_{k=1}^{n} a_{jk}\, c_j\, d_k$$

where c_j and d_k are characteristic functions of partitions S and T into sets of equal measure.

PROOF. By 1.52, there exist $x_1, \ldots, x_n \in L_p(S)$ and $y_1, \ldots, y_n \in L_p(T)$ such that

$$\left\| z - \sum_{i=1}^{n} x_i \otimes y_i \right\|_p < \epsilon/2.$$

124

By 10.14, there exist simple functions $g_1, \ldots, g_n$ in $L_p(S)$ such that

$$\|y_i\|_p \|x_i - g_i\|_p < \epsilon/4n$$

and each g_i has a representation

$$g_i(s) = \sum_{j=1}^{m} \lambda_{ij}\, c_j(s)$$

in which $c_j c_k = 0$ for $j \neq k$, $\Sigma c_j = 1$, and $\int c_j(s)ds = \mu(S)/m$. Similarly, we find simple functions $h_i \in L_p(T)$ such that

$$\|g_i\|_p \|y_i - h_i\|_p < \epsilon/4n$$

and each h_i has a representation

$$h_i(t) = \sum_{k=1}^{r} \alpha_{ik}\, d_k$$

where the characteristic functions d_k have the same properties as the c_j. We put

$$f = \sum_{i=1}^{n} g_i \otimes h_i.$$

Then

$$\|z - f\|_p \leq \|z - \Sigma x_i \otimes y_i\|_p + \|\Sigma x_i \otimes y_i - \Sigma g_i \otimes y_i\|_p + \|\Sigma g_i \otimes y_i - \Sigma g_i \otimes h_i\|_p$$

$$\leq \epsilon/2 + \Sigma\|x_i - g_i\|_p \|y_i\|_p + \Sigma\|g_i\|_p \|y_i - h_i\|_p < \epsilon.$$

Now the form of f is

$$f = \sum_{i=1}^{n} \left(\sum_{j=1}^{m} \lambda_{ij}\, c_j \right) \otimes \left(\sum_{k=1}^{r} \alpha_{ik}\, d_k \right)$$

$$= \sum_{j=1}^{m} \sum_{k=1}^{r} \left(\sum_{i=1}^{n} \lambda_{ij}\, \alpha_{ik} \right) c_j\, d_k. \quad \blacksquare$$

125

CHAPTER 11

APPENDIX ON MISCELLANEOUS RESULTS IN BANACH SPACES

11.1 LEMMA. *If* $P : X \twoheadrightarrow U$ *and* $Q : X \twoheadrightarrow V$ *are Banach space projections such that* $PQP = QP$, *then the Boolean sum*

$$P \oplus Q = P + Q - PQ$$

is a projection of X *onto* $U + V$. *The latter is therefore closed.*

PROOF. It is evident that $P \oplus Q$ maps X into $U + V$. In order to verify that it leaves invariant each element of U, it suffices to write

$$(P \oplus Q)P = P^2 + QP - PQP = P + QP - QP = P.$$

Similarly, $P \oplus Q$ leaves invariant each element of V, since

$$(P \oplus Q)Q = PQ + Q^2 - PQ^2 = PQ + Q - PQ = Q. \quad \blacksquare$$

11.2 THEOREM. *Let* P *be a projection of a Banach space* X *onto a subspace* G, *and* Q *a projection of a Banach space* Y *onto a subspace* H. *Let* α *be a uniform norm on* $X \otimes Y$. *Then*

$$(P \otimes_\alpha I) \oplus (I \otimes_\alpha Q)$$

is a projection of $X \otimes_\alpha Y$ *onto*

$$G \bar{\otimes} Y + X \bar{\otimes} H.$$

The latter is therefore closed in $X \otimes_\alpha Y$.

PROOF. The operator $P \otimes I$ is a projection of $X \otimes Y$ onto $G \otimes Y$. By the uniformity of α, $\|P \otimes I\| = \|P\|$. The continuous extension, $P \otimes_\alpha I$, is a projection of $X \otimes_\alpha Y$ onto $G \bar{\otimes} Y$. Similarly $I \otimes_\alpha Q$ is a projection of $X \otimes_\alpha Y$ onto $X \bar{\otimes} H$. By using dyads, linearity, and continuity, we can verify that

$$(P \otimes_\alpha I)(I \otimes_\alpha Q) = (I \otimes_\alpha Q)(P \otimes_\alpha I).$$

The result then follows from 11.1. $\quad \blacksquare$

11.3 THEOREM. *For a pair of closed subspaces U and V in a Banach space, these six properties are equivalent:*

(i) $U + V$ *is closed.*

(ii) *There is a constant c such that each $w \in U + V$ can be represented as $w = u + v$, with $u \in U$, $v \in V$, and $\|u\| + \|v\| \le c\|w\|$.*

(iii) $U^{\perp} + V^{\perp} = (U \cap V)^{\perp}$.

(iv) $U^{\perp} + V^{\perp}$ *is weak*-closed.*

(v) $U^{\perp} + V^{\perp}$ *is proximinal.*

(vi) $U^{\perp} + V^{\perp}$ *is closed.*

PROOF. (i) implies (ii) [**149**, p. 130]. Introduce a norm on $U \times V$ by writing $\|(u, v)\| = \|u\| + \|v\|$. Define a linear map $L : U \times V \to U + V$ by writing $L(u, v) = u + v$. The map L is continuous and surjective. Since $U + V$ is complete, the open-mapping theorem applies [**149**, p. 48]. Hence for some c and for each $w \in U + V$ there is an element $(u, v) \in U \times V$ such that $L(u, v) = w$ and $\|(u, v)\| \le c\|w\|$.

(ii) implies (iii) [**108**, p. 390]. The inclusion $U^{\perp} + V^{\perp} \subset (U \cap V)^{\perp}$ is elementary. For the reverse inclusion, let $\varphi \in (U \cap V)^{\perp}$. Define ψ on $U + V$ as follows. If $w \in U + V$, write $w = u + v$ with $u \in U$, $v \in V$, and $\|u\| + \|v\| \le c\|w\|$. Put $\psi(w) = \varphi(v)$. The definition is proper, for if $w = u' + v'$ then $u' - u = v - v'$, whence $v - v' \in U \cap V$ and $\varphi(v - v') = 0$. Also, ψ is continuous since

$$|\psi(w)| = |\varphi(v)| \le \|\varphi\| \, \|v\| \le c\|\varphi\| \, \|w\|.$$

Let X be the Banach space in which U and V are situated. By the Hahn-Banach Theorem, there exists $\theta \in X^*$ such that $\theta \mid (U + V) = \psi$. Then $\theta \in U^{\perp}$ since

$$\theta(u) = \psi(u) = \psi(u + 0) = \phi(0).$$

If $v \in V$ then

$$(\varphi - \theta)(v) = \varphi(v) - \psi(v) = \varphi(v) - \varphi(v) = 0.$$

Hence $\varphi - \theta \in V^{\perp}$, and $\varphi = \theta + (\varphi - \theta) \in U^{\perp} + V^{\perp}$.

(iii) $\Longrightarrow$ (iv). The annihilator of any subspace of X is weak*-closed in X^*.

(iv) $\Longrightarrow$ (v). Any weak*-closed set in X^* is proximinal [**104**, p. 123].

(v) $\Longrightarrow$ (vi). Any proximinal set is closed.

(vi) $\Longrightarrow$ (i) [**144**, p. 555]. Define a linear map $L : U \to X/V$ by writing $Lu = u + V$. Let

$$Q : X \to X/V \quad \text{and} \quad Q' : X^* \to X^*/U^{\perp}$$

be the quotient maps. Let

$$A : (X/V)^* \to V^\perp \quad \text{and} \quad B : X^*/U^\perp \to U^*$$

be the canonical isometries given by $A\psi(x) = \psi(x+V)$ and $B(\varphi+U^\perp) = \varphi \mid U$. One can then verify that $L^* = BQ'A$ and that the range of L^* is $B[U^\perp + V^\perp]$. Since $U^\perp + V^\perp$ is assumed closed, so is the range of L^*. By [**57**, p. 488], the range of L is closed. This last space is $U+V$, as a subspace of X/V. Finally, $U+V$ is closed in X since it is $Q^{-1}[U+V]$. ∎

11.4 LEMMA. *Let U and V be weak*-closed subspaces in a conjugate Banach space X^*. If $U + V$ is norm-closed, then it is weak*-closed and proximinal.*

PROOF. Since U and V are weak*-closed, they obey the equations $U = (U_\perp)^\perp$ and $V = (V_\perp)^\perp$ [**108**, p. 233]. Here we have put

$$U_\perp = \{x \in X : \phi(x) = 0 \quad \text{for all} \quad \phi \in U\}.$$

By 11.3, the desired conclusion follows. ∎

11.5 LEMMA. *Let G be a finite-dimensional subspace with basis $\{g_1, \ldots, g_n\}$ in a Banach space X. Let Y be another Banach space, and α a reasonable crossnorm on $X \otimes Y$. Then there exist bounded linear maps $L_i : X \otimes_\alpha Y \to Y$ such that*

$$z = \sum_{i=1}^n g_i \otimes L_i z \quad \text{for all} \quad z \in G \otimes Y.$$

PROOF. Select $\varphi_1, \ldots, \varphi_n \in X^*$ so that $\varphi_i(g_j) = \delta_{ij}$. Define L_i on the (uncompleted) tensor product $X \otimes Y$ by

$$L_i\left(\sum_j x_j \otimes y_j \right) = \sum_j \varphi_i(x_j) y_j \quad (1 \le i \le n).$$

Notice that if the element $z = \sum_j x_j \otimes y_j$ is interpreted as a member of $\mathcal{L}(X^*, Y)$, the defining equation for L_i says, in effect, $L_i z = z(\varphi_i)$. Thus the definition of $L_i z$ is independent of the representation of z. Since $\lambda(z)$ is the norm of z as an operator, we have, by 1.6,

$$\|L_i z\| = \|z(\varphi_i)\| \le \lambda(z)\|\varphi_i\| \le \alpha(z)\|\varphi_i\|.$$

This shows that

$$\|L_i\| = \sup\{\|L_i z\| : \alpha(z) \le 1\} \le \|\varphi_i\|.$$

The operator L_i is extended by continuity to $X \otimes_\alpha Y$, with the same bound [**165**, p. 39]. If $z \in G \otimes Y$ then

$$z = \sum_{j=1}^n g_j \otimes y_j$$

128

for appropriate $y_j \in Y$. Hence

$$L_i z = z(\varphi_i) = \sum_{j=1}^{n} \varphi_i(g_j) y_j = y_i. \quad \blacksquare$$

11.6 LEMMA. *Let G and H be finite-dimensional subspaces in Banach spaces X and Y respectively. Let $\{g_1, \ldots, g_n\}$ and $\{h_1, \ldots, h_m\}$ be bases for G and H respectively. Let α be a reasonable crossnorm on $X \otimes Y$. Then there exist bounded linear operators*

$$\mathcal{G}_i : X \otimes_\alpha Y \to Y \quad \text{and} \quad \mathcal{H}_j : X \otimes_\alpha Y \to X$$

such that for each $w \in W = G \otimes Y + X \otimes H$,

$$w = \sum_{i=1}^{n} g_i \otimes \mathcal{G}_i w + \sum_{j=1}^{m} \mathcal{H}_j w \otimes h_j.$$

PROOF. By 11.2, there exist projections

$$P : X \otimes_\alpha Y \twoheadrightarrow G \otimes Y \quad \text{and} \quad Q : X \otimes_\alpha Y \twoheadrightarrow X \otimes H$$

such that $P + Q - QP$ is a projection of $X \otimes_\alpha Y$ onto W.

By 11.5 there exist bounded linear maps

$$L_i : X \otimes_\alpha Y \to Y \quad \text{and} \quad K_j : X \otimes_\alpha Y \to X$$

such that

$$z = \sum_{i=1}^{n} g_i \otimes L_i z \qquad (z \in G \otimes Y)$$

$$z = \sum_{j=1}^{m} K_j z \otimes h_j \qquad (z \in X \otimes H).$$

Define $\mathcal{G}_i = L_i P$ and $\mathcal{H}_j = K_j Q(I - P)$. Then for any $w \in W$,

$$
\begin{aligned}
w &= Pw + Q(I - P)w \\
&= \sum_{i=1}^{n} g_i \otimes L_i Pw + \sum_{j=1}^{m} K_j Q(I - P)w \otimes h_j \\
&= \sum_{i=1}^{n} g_i \otimes \mathcal{G}_i w + \sum_{j=1}^{m} \mathcal{H}_j w \otimes h_j. \quad \blacksquare
\end{aligned}
$$

11.7 LEMMA. *If $f \in C(S \times T)$ then the map $s \mapsto f_s$ from S to $C(T)$ is continuous. Hence the set $\{f_s : s \in S\}$ is equicontinuous in $C(T)$.*

PROOF. Let s_0 be any point of S. Given $\epsilon > 0$, we will find a neighborhood $\mathcal{N}$ of s_0 such that $\|f_s - f_{s_0}\| < \epsilon$ whenever $s \in \mathcal{N}$. Thus we must determine $\mathcal{N}$ so that

$$|f^t(s) - f^t(s_0)| < \epsilon \qquad (s \in \mathcal{N}, \, t \in T).$$

Since f is continuous at each point (s_0, t), there exist neighborhoods $U(t)$ of s_0 and $V(t)$ of t such that

$$|f(\sigma, \tau) - f(s_0, t)| < \epsilon/2 \quad \text{whenever} \quad (\sigma, \tau) \in U(t) \times V(t).$$

By compactness of T, we find $t_1, \ldots, t_n$ such that $V(t_1), \ldots, V(t_n)$ cover T. Define

$$\mathcal{N} = \bigcap_{j=1}^{n} U(t_j).$$

This neighborhood of s_0 has the property desired, for if $s \in \mathcal{N}$, if $t \in T$, and if $t \in V(t_i)$, then (s, t) and (s_0, t) belong to $U(t_i) \times V(t_i)$, whence

$$|f^t(s) - f^t(s_0)| \leq |f(s, t) - f(s_0, t_i)| + |f(s_0, t_i) - f(s_0, t)| < \epsilon.$$

Since the map $s \mapsto f_s$ is continuous and S is compact, the set $\{f_s : s \in S\}$ is compact in $C(T)$. Hence it is closed, bounded, and equicontinuous by the Ascoli Theorem. ∎

11.8 LEMMA. *Let f be a continuous map of a Banach space X into a Banach space Y. If T is a compact Hausdorff space then the map defined by $\Phi(z) = f \circ z$ is continuous from $C(T, X)$ to $C(T, Y)$.*

PROOF. It is obvious that Φ maps $C(T, X)$ into $C(T, Y)$. For the continuity of this map we proceed by contradiction. Suppose that there is a net $z_\alpha \to z_0$ such that

$$\|f \circ z_\alpha - f \circ z_0\| \geq \epsilon.$$

For each α, select $t_\alpha \in T$ such that

$$\|f(z_\alpha(t_\alpha)) - f(z_0(t_\alpha))\| \geq \epsilon.$$

Take a convergent subnet, $t_\beta \to t_0$. By the continuity of f at $z_0(t_0)$, there exists $\delta > 0$ such that for all $x \in X$,

$$\|x - z_0(t_0)\| < \delta \implies \|f(x) - f(z_0(t_0))\| < \epsilon/2.$$

By the continuity of z_0 at t_0, there is an index β such that

$$\|z_0(t_\beta) - z_0(t_0)\| < \delta/2 \quad \text{and} \quad \|z_\beta - z_0\| < \delta/2.$$

Then

$$\|z_\beta(t_\beta) - z_0(t_0)\| \leq \|z_\beta(t_\beta) - z_0(t_\beta)\| + \|z_0(t_\beta) - z_0(t_0)\| < \frac{\delta}{2} + \frac{\delta}{2}.$$

Hence

$$\|f(z_\beta(t_\beta)) - f(z_0(t_0))\| < \epsilon/2 \quad \text{and} \quad \|f(z_0(t_\beta)) - f(z_0(t_0))\| < \epsilon/2.$$

It follows from these two inequalities that

$$\|f(z_\beta(t_\beta)) - f(z_0(t_\beta))\| < \epsilon,$$

which is a contradiction. ■

11.9 COROLLARY. *If f is a continuous map of $C(S)$ into itself, then the map $\bar{f}$ defined by $(\bar{f}z)(s,t) = (f(z^t))(s)$ is continuous from $C(S \times T)$ into itself.*

PROOF. Let J be the canonical isometry of $C(S \times T)$ onto $C(T, C(S))$ defined by

$$(Jz)(t) = z^t \qquad (z \in C(S \times T)).$$

Let Φ be the mapping of $C(T, C(S))$ to $C(T, C(S))$ defined by

$$\Phi v = f \circ v \qquad v \in C(T, C(S)).$$

By 11.8, Φ is continuous. Since $\bar{f} = J^{-1}\Phi J$, $\bar{f}$ is continuous. ■

11.10 LEMMA. *Let S be a compact Hausdorff space. Let μ be a regular Borel measure such that $\mu(S) < \infty$. If $w \in L_1(S)$ and $\int xw = 0$ for all $x \in C(S)$ then $w = 0$.*

PROOF. Since sgn $w \in L_\infty(S)$, we can find by Lusin's Theorem [**148**, p. 56] a sequence $\{x_n\}$ in $C(S)$ such that $\|x_n\|_\infty \leq 1$ and if

$$A_n = \{s : x_n(s) \neq \text{sgn } w(s)\}$$

then $\mu(A_n) \to 0$. Now

$$\|w\|_1 = \int |w| = \int w \text{ sgn } w = \int w x_n + \int w(\text{sgn } w - x_n)$$
$$= \int_{A_n} w(\text{sgn } w - x_n) \leq 2 \int_{A_n} |w|.$$

Since $\mu(A_n) \to 0$, it follows from 10.10 that $\int_{A_n} |w| \to 0$ and so $w = 0$. ■

11.11 LEMMA. *(Auerbach) If X is an n-dimensional Banach space, then there exist $x_1, \ldots, x_n$ in X and $\phi_1, \ldots, \phi_n$ in X^* such that*

$$\|x_i\| = \|\phi_i\| = 1 \text{ and } \phi_i(x_j) = \delta_{ij} \text{ for } 1 \leq i, j \leq n.$$

PROOF. Fix any convenient basis $\{y_i, \ldots, y_n\}$ for X. Select $\phi_1, \ldots, \phi_n$ in the unit cell of X^* to maximize the determinant of the matrix A whose elements are $A_{ij} = \phi_i(y_j)$. Let B denote the matrix inverse of A, and define $x_j = \sum_{\nu=1}^n B_{\nu j} y_\nu$. Then

$$\phi_i(x_j) = \phi_i\left(\sum_{\nu=1}^n B_{\nu j} y_\nu\right) = \sum_{\nu=1}^n B_{\nu j}\, \phi_i(y_\nu) = \sum_{\nu=1}^n B_{\nu j}\, A_{i\nu} = \delta_{ij}.$$

If $\lambda = \det(A)$ then the cofactor of A_{ij} is λB_{ji}. Let ψ be a functional of norm 1. If we substitute ψ for ϕ_i in the definition of A, the determinant is not increased. Computing this determinant, using the elements in the ith row and their cofactors, gives us

$$\sum_{\nu=1}^n \psi(y_\nu)\lambda B_{\nu i} \leq \lambda.$$

It follows that $\psi(x_i) \leq 1$. Since ψ was arbitrary, $\|x_i\| \leq 1$. Since $\phi_i(x_i) = 1$, $\|x_i\| = 1$. ∎

11.12 THEOREM. *Let A be a compact linear operator from a Banach space X into a Banach space Y. Let J_Y be the canonical embedding of Y into Y^{**}. Then $A^{**}(X^{**}) \subset J_Y(Y)$.*

PROOF. We assert that if X^{**} and Y^{**} are given their weak*-topologies then A^{**} is a continuous mapping from X^{**} into Y^{**}. Indeed, if $\{p_\alpha\}$ is a net in X^{**} with weak*-limit 0 then, for $\psi \in Y^*$, we have

$$\lim_\alpha (A^{**}p_\alpha)(\psi) = \lim_\alpha p_\alpha(A^*\psi) = 0.$$

Now take $p \in X^{**}$. By the Goldstine Theorem [57, p. 424] we can find a net $\{p_\alpha\}$ in $J_X(X)$ whose weak*-limit is p and $\|p_\alpha\| \leq \|p\|$. Now for any $\psi \in Y^*$, we have

$$(A^{**}p)(\psi) = \lim_\alpha (A^{**}p_\alpha)(\psi) = \lim_\alpha \psi(A J_X^{-1} p_\alpha).$$

By the compactness of A, $\{A J_X^{-1} p_\alpha\}$ contains a subnet $\{A x_\beta\}$ which converges in Y to an element y. Hence

$$(A^{**}p)(\psi) = \lim_\beta \psi(A x_\beta) = \psi(y) = (J_Y y)(\psi)$$

and so $A^{**}p = J_Y(y)$. ∎

11.13 DEFINITION. *Let Φ be a set-valued mapping, taking points of a topological space S into the family of all subsets of a topological space T. The mapping Φ is said to be lower semicontinuous if, for each open set $\mathcal{O}$ in T, the set*

$$\{s \in S : \Phi(s) \cap \mathcal{O} \neq \square\}$$

is open in S.

11.14 THEOREM. *(Michael Selection Theorem)* *Let Φ be a lower semicontinuous map of a paracompact space S into the family of nonvoid closed convex subsets of a Banach space X. Then Φ has a continuous selection; i.e., there exists a continuous map $\varphi : S \to X$ such that $\varphi(s) \in \Phi(s)$ for all $s \in S$.*

11.15 DEFINITION. *Let Φ be a set-valued mapping, taking each point of a measurable space S into a subset of a topological space T. We say that Φ is weakly measurable if $\Phi^{-}(\mathcal{O})$ is measurable in S whenever $\mathcal{O}$ is open in T. Here we have put, for any $A \subset T$,*

$$\Phi^{-}(A) = \{s \in S : \Phi(s) \cap A \neq \square\}.$$

11.16 THEOREM. *(Kuratowski-Ryll-Nardzewski Measurable Selection Theorem)* *Let Φ be a weakly measurable set-valued map which carries each point of a measurable space S to a closed nonvoid subset of a complete separable metric space. Then Φ has a measurable selection; i.e., there exists a function $f : S \to T$ such that $f(s) \in \Phi(s)$ for each $s \in S$ and $f^{-1}(\mathcal{O})$ is measurable for each open set $\mathcal{O}$ in T.*

11.17 THEOREM. *Let Φ be a set-valued map, carrying each point of a measurable space S to a closed nonvoid subset of a finite-dimensional Banach space X. If $\Phi^{-}(K)$ is measurable for each compact K in X then Φ has a measurable selection.*

PROOF. Let $\{x_1, x_2, \ldots\}$ be a countable dense set in X. Consider the family of closed cells

$$C_{nm} = \{x : \|x - x_n\| \leq 1/m\} \qquad (n, m \in \mathbb{N}).$$

It is easily seen that each open set $\mathcal{O}$ in X can be expressed as

$$\mathcal{O} = \cup\{C_{nm} : C_{nm} \subset \mathcal{O}\}.$$

Since each C_{nm} is compact, $\Phi^{-}(C_{nm})$ is measurable. Hence $\Phi^{-}(\mathcal{O})$ is measurable, since

$$\Phi^{-}(\mathcal{O}) = \cup\{\Phi^{-}(C_{nm}) : C_{nm} \subset \mathcal{O}\}.$$

This proves that Φ is weakly measurable. An application of 11.16 completes the proof. ∎

NOTES AND REMARKS

CHAPTER 1. The material of Chapter 1 is accessible in a number of sources. We refer the reader to various monographs and textbooks rather than to original sources. An early, but very readable, treatment is Schatten's monograph [154]. Chapter 9 in Diestel and Uhl [55] is also helpful. Among textbooks which treat the tensor product of linear topological spaces, are those of Treves [165], Day [41], Köthe [113], and Schaefer [153]. The article by Gilbert and Leih [67] contains a succinct account of tensor-product theory.

The norms α_p discussed in 1.43 to 1.46 were introduced independently by Saphar [152] and Chevet [36]. The paper of Saphar [151] contains a wealth of information about them. The proof of the triangle inequality in 1.46 is from [151]. The isomorphism theorem 1.52 was apparently obtained independently by Chevet [35] and Persson [140].

CHAPTER 2. The first half of Chapter 2 deals with the proximinality of $G \otimes Y$ in $X \otimes_\alpha Y$, where X and Y are Banach spaces, G is a subspace of X, and α is a crossnorm. Results 2.1 to 2.3 and 2.5 to 2.7 are from Franchetti and Cheney [61]. The distance formula in $C(S, Y)$ (2.4) was given, along with many other important results, by Buck [25]. The part of 2.4 which refers to $\ell_\infty(S, Y)$ is from von Golitschek and Cheney [76]. The results 2.10 to 2.13, concerning proximinality in spaces $L_1(S, Y)$, are contained implicitly or explicitly in Khalil [112]. The result 2.13 generalizes work in [121], where techniques like those in 2.8 gave a weaker result. An alternative approach to some of these proximinality questions is via the theory of Chebyshev centers. These ideas are explored in Franchetti and Cheney [62]. For recent results in the abstract theory of Chebyshev centers, the paper of Amir, Mach, and Saatkamp [1] is recommended.

In the second half of Chapter 2, the theme is the proximinality of subspaces $X \bar{\otimes} H + G \bar{\otimes} Y$ in $X \otimes_\alpha Y$. Results 2.14 to 2.18 are from Respess and Cheney [146]. Corollary 2.19 was given first in [121]. Results 2.21 to 2.25 are also from [146]. Theorem 2.22 has been called the "Sitting Duck Theorem" because it seems likely that its hypotheses can be drastically weakened. The situation in $L_1(S \times T)$ is much more satisfactory, as can be seen by comparing 2.22 with 2.26. The latter is from Holland, Light, and Sulley [102].

The problem of characterizing best approximations in spaces $G \otimes Y + X \otimes H$ has not received much attention. The general theorems of Singer [160] are of course applicable. Havinson [99] proved a characterization theorem for best approximations in $C(S \times T)$ by elements of the subspace $C(S) + C(T)$. In [120], there is a characterization theorem similar to Havinson's for subspaces $C(S) \otimes H + G \otimes C(T)$, where G and H are finite dimensional.

Approximation in $C(D)$, where D is a proper subset of $S \times T$, differs in essential aspects from appproximation in $S \times T$. As 2.27 shows, the proximinality of $C(S) + C(T)$ in $C(D)$ depends on the geometry of D. The reader should consult the papers of Ofman [137] and von Golitschek [69] for further information.

Theorem 2.22 requires a proximity map of $C(S)$ satisfying a Lipschitz condition. Unfortunately, this is a rather restrictive hypothesis, as the subspaces which are usually encountered in practical problems do not have such proximity maps. For example, Haar subspaces of dimension 2 or greater never have this property. Indeed, subspaces having the Haar property at a single point do not have Lipschitz proximity maps. See Respess and Cheney [145]. One can further prove that the proximity map onto a Haar subspace of dimension 2 or greater is not uniformly continuous on bounded sets. By a theorem of Freud, the proximity map of $C(S)$ onto a Haar subspace is *locally* Lipschitz, but this property seems to be inadequate for the proof of theorem 2.22. In general, $C(S)$ will contain subspaces of any dimension (finite or infinite) which possess Lipschitz proximity maps [145].

The proximinality of spaces $C(S) \otimes \Pi_m(T) + \Pi_n(S) \otimes C(T)$, where S and T are intervals and Π_n is the space of polynomials of degree at most n, is an open question, except when $\min(n, m)$ is 0. Von Golitschek and Cheney in [76] prove that each function in $C(S \times T)$ possesses a best approximation in $\ell_\infty(S) \otimes \Pi_1(T) + \Pi_1(S) \otimes \ell_\infty(T)$ which is continuous on the *interior* of the rectangle $S \times T$.

CHAPTER 3. According to Deutsch [51], the alternating algorithm appeared first in mimeographed lecture notes of von Neumann in 1933. These notes were later published in [135]. References to subsequent work on this algorithm for Hilbert space projections will be found in Deutsch [49]. Applications of the algorithm to diverse problems, such as the multigrid method for partial differential equations and to computerized tomography, are mentioned in Deutsch [48]. Result 3.4 is from Franchetti and Light [63]. Item 3.5 is from [51]. The theorem in 3.6 derives from a paper by Baillon, Bruck and Reich [19]; item 3.7 is from [63]. The theorem in 3.8 is of course von Neumann's original result [135]. Results 3.9 and 3.10 are from [51]; the sequence 3.11 to 3.16, from [63].

Franchetti and Light in [63] have described a class of uniformly convex Banach spaces in which the alternating algorithm is effective for any pair of closed subspaces. This class properly includes the Hilbert spaces.

CHAPTER 4. The concept of a central proximity map comes from Golomb's work [78]. Items 4.1, 4.2, and 4.3 are from [78], although the ideas are present in Diliberto and Straus [56]. Item 4.7 is also from [78].

CHAPTER 5. Lemmas 5.1 and 5.2 are implicit in Diliberto and Straus [56]. Items

5.3 and 5.4 are from Aumann [6]. Item 5.6 comes from [56], but the proof is due to Aumann [6]. The result 5.8 derives from [6], but the proof is from Light and Cheney [122].

It was proved by Dyn [58] that the Diliberto-Straus algorithm will fail if applied to the subspaces $C(S) \otimes \Pi_1(T)$ and $\Pi_0(S) \otimes C(T)$. Her method of proof was used by von Golitschek and Cheney [75] to prove that the algorithm fails for $C(S) \otimes H$ and $G \otimes C(T)$ whenever G and H are Haar subspaces such that dim $G > 1$ and dim $H > 0$. There exist non-Haar subspaces G and H of arbitrary dimension, however, for which the algorithm does work. See Respess and Cheney [145].

The convergence of the Diliberto-Straus algorithm can be arbitrarily slow, as is shown by examples in von Golitschek and Cheney [77]. An application of the algorithm to the scaling (or "preconditioning") of matrices in the numerical solution of partial differential equations can be found in several papers of Bank [20,21]. Often, one or two steps of the algorithm suffice to produce a satisfactory preconditioning. The theoretical basis for this is discussed in [77], where it is shown, for example, that the third and fourth iterates in the algorithm satisfy the inequalities $\|z_3\| \leq 2 \operatorname{dist}(z, W)$ and $\|z_4\| \leq 1.5 \operatorname{dist}(z, W)$.

CHAPTER 6. The results in this chapter are due to von Golitschek in a series of papers [69-74], especially [69]. For simplicity of exposition, we have limited the discussion to domains $S \times T$, although the original work includes certain types of closed subsets in $S \times T$.

CHAPTER 7. The results 7.1 to 7.12 derive from Light and Holland [123]. Example 7.13 is from Light, McCabe, Phillips, and Cheney [124]. Lemma 7.15 comes from [124]. Items 7.16 to 7.19 are from [117] and [118].

CHAPTER 8. Result 8.2 is due to Jameson and Pinkus [109]. Theorem 8.8 is from Halton and Light [89]. Results 8.13 and 8.14 come from Franchetti and Cheney [60], and are based upon the work in [109]. We thank Dr. Shekhtman for communicating Example 8.15 to us.

CHAPTER 9. Theorem 9.1 is due to Rudin [150]. The techniques used in 9.2 were kindly communicated to us by Y. Benyamini. Corollary 9.4 is from Halton and Light [90]. Theorem 9.5 comes from Halton and Light [88], while results 9.7 to 9.9 are from [90].

It is possible to obtain Theorem 9.9 directly from 9.1 without using the discretization involved in 9.2 to 9.8. Details of this approach are contained in a preprint by W.A.Light entitled "Minimal Projections in Tensor Product Spaces".

CHAPTER 10. The material on the Bochner integral can be found in Diestel and Uhl [55].

CHAPTER 11. Lemma 11.1 on the Boolean sum of two projections is probably due to Gordon [81]; see also Gordon and Cheney [84]. Theorem 11.2 is from Respess and Cheney [146]. In Theorem 11.3, the equivalence of (i) and (ii) occurs in Rudin [149, p. 130]. The equivalence of (i) and (vi) is due to Reiter [144]. The proof of Auerbach's Lemma is given by A.F. Ruston in "Auerbach's lemma and tensor products of Banach spaces," Proc. Cambridge Phil. Soc. **58** (1964), 476-480. The Michael Selection Theorem appeared first in Michael [130]. See also his [131], Holmes' book [103], and Parthasarathy's survey [139]. The Kuratowski-Ryll-Nardzewski Measurable Selection Theorem first appeared in their paper [114]. See also [139], Wagner's survey update [166], and Diestel's book [54, p. 268].

BIOGRAPHICAL SKETCH OF ROBERT SCHATTEN

Robert Schatten was born in Lwów, Poland, on January 28, 1911. He received the Magister degree from John Casimir University (Lwów) in 1933. After emigrating to the United States, he enrolled in the graduate school of Columbia University, receiving an M.A. degree in 1939. He continued his research under the direction of Francis J. Murray, and was awarded the Ph.D. degree in 1942. He held a brief appointment as a lecturer in the College of Pharmacy at Columbia University in 1942 before joining the U.S. Army, in which he served from 1942 to 1943. He suffered a broken back during training at Fort Benning, Georgia, and this injury gave him much pain for the rest of his life. In the academic year 1943-1944, he had an appointment as assistant professor at the University of Vermont. He then won a two-year appointment as a Fellow of the National Research Council, and divided his time during this period between the Institute for Advanced Study and Yale University. He collaborated during these years with John von Neumann and with Nelson Dunford. In 1946, Schatten began a long association with the University of Kansas, first as associate professor (1946-1952) and then as professor (1952-1961). This tenure was interrupted by leaves in 1950 and in 1952-1953, both of which he spent at the Institute for Advanced Study. The year 1960-1961 was spent as a visiting professor at the University of Southern California, and in 1961-1962 he served as professor at the State University of New York at Stony Brook. In 1962 he became professor at Hunter College, where he remained until his death on August 26, 1977. During the years 1964-1972 he was also a member of the doctoral faculty of the Graduate School of the City University of New York. At the time of his death there were no immediate survivors, all his known relatives in Poland having been killed during the war.

To his former students, Schatten will be remembered as a dedicated teacher who was genuinely concerned with the intellectual development of his students. They will certainly not forget his unique style of lecturing. He always spoke without a book or notes, and rarely used the blackboard. His lectures were extremely clear and well-organized; he never lost his way in complicated arguments. The pace was such that the students could (and were expected to) take notes verbatim; if they did so, their notes would read like a polished book, except for some linguistic idiosyncracies such as, "Given is a set...". He left nothing to chance in his dictation; for example, he invariably ended an argument with "This concludes the proof."

Schatten had his own way of making abstract concepts memorable to his elementary classes. Who could forget what a *sequence* was after hearing Schatten describe a long corridor, stretching as far as the eye could see, with hooks regularly spaced on the wall and numbered $1, 2, 3, \ldots$? "Then," Schatten would say, "I come along with a big bag of numbers over my shoulder, and hang one number on each hook." This of course was accompanied by suitable gestures for emphasis.

Schatten had some eccentricities which endeared him to his friends. He hated noise, especially when it interrupted his sleep. In Lawrence, Kansas, he was seen early one morning in his garden, clad in pyjamas, trying to shoo away the grackles from a tree near his bedroom. Cars were also his bêtes noires: although he owned a car at one time, he never fully mastered the art of driving. He once got a nasty bruise from attempting to put his head out of the car window before lowering the glass. Bachelor life also presented various pitfalls such as having to contend with laundries that insisted on ironing his socks. He kept his unpublished mathematical researches in a bank's safe-deposit box.

Schatten's principal mathematical achievement was that of initiating the study of tensor products of Banach spaces. The concepts of crossnorm, associate norm, greatest crossnorm, least crossnorm, and uniform crossnorm, all either originated with him or at least first received careful study in his papers. He was mainly interested in the applications of this subject to linear transformations on Hilbert space. In this subject, the "Schatten Classes" perpetuate his name.

Publications of Robert Schatten

1. *On the direct product of Banach spaces,* Trans. Amer. Math. Soc. **53** (1943), 195-217. MR **4**-161.

2. *On reflexive norms for the direct product of Banach spaces,* Trans. Amer. Math. Soc. **54** (1943), 498-506. MR **5**-99.

3. (with N. Dunford), *On the associate and conjugate space for the direct product of Banach spaces,* Trans. Amer. Math. Soc. **59** (1946), 430-436. MR **7**-455.

4. *The cross space of linear transformations,* Ann. of Math. (2) **47** (1946), 73-84. MR **7**-455.

5. (with J. von Neumann), *The cross-space of linear transformations. II,* Ann. of Math. (2) **47** (1946), 608-630. MR **8**-31.

6. (with J. von Neumann), *The cross-space of linear transformations. III,* Ann. of Math. (2) **49** (1948), 557-582. MR **10**-256.

7. *On projections with bound* 1, Ann. of Math. (2) **48** (1947), 321-325. MR **10**-128.

8. "A Theory of Cross-Spaces," Annals of Mathematics Studies, **no. 26**, Princeton University Press, Princeton, N.J., 1950. MR **12**-186.

9. *"Closing-up" of sequence spaces,* Amer. Math. Monthly **57** (1950), 603-616. MR **12**-418.

10. *The space of completely continuous operators on a Hilbert space,* Math. Ann. **134** (1957), 47-49. MR **19**-756.

11. "Norm Ideals of Completely Continuous Operators," Ergebnisse der Mathematik und ihre Grenzgebiete, N.F. Heft **27** Springer-Verlag, Berlin, 1960. MR **22** #9878 and MR **41** #2449. Reviewed in Bull. Amer. Math. Soc.**7** (1961),532.

12. *A non-approximation theorem,* Amer. Math. Monthly 9 (1962), 745-750.

13. *On the trace-class of operators, I,* National Bureau of Standards Report 7580 (July, 1962), 22pp. Available from Record Group 167, Records of the National Bureau of Standards, National Archives, Washington, D.C. 20408.

14. *The greatest cross norm,* J. Res. Nat. Bur. Standards Sect. B, **68B** (1964), 185-193. MR **31** #3875.

15. *A remark on the approximation problem,* Annali di Mat. Pura ed Applicata (4) **98** (1974), 235-238. MR **50** #5431.

Sources

"American Men Of Science, The Physical and Biological Sciences, 11th edition." R.R. Bowker Co., New York, 1967, 4692.

Obituary, New York Times, August 30, 1977, 32.

Mathematical Reviews.

Personal communications from Professor Emeritus G. Baley Price (University of Kansas), Professor Richard Churchill (Hunter College), Professor Peter Falley (Fairleigh Dickinson University), and Professor Charles D. Masiello (Pace University, Pleasantville, New York).

REFERENCES

1. D. Amir, J. Mach, and K. Saatkamp, *Existence of Chebyshev centers, best n-nets and best compact approximants*, Trans. Amer. Math. Soc. **271** (1982), 513-524.

2. B. Atlestam and F. Sullivan, *Iteration with best approximation operators*, Rev. Roum. Math. Pures Appl. **21** (1976), 125-131. MR **53** #6188.

3. G. Aumann, *Approximation von Funktionen*, in "Mathematische Hilfsmittel des Ingenieurs, Teil III," R. Sauer and I. Szabó, eds., Springer-Verlag, Berlin, 1968, 320-351.

4. —————————, *Approximation by step functions*, Proc. Amer. Math. Soc. **14** (1963), 477-482. MR **27** #515.

5. —————————, *Lineare Approximationen auf einem Geflecht*, Arch. Math. **10** (1959), 267-272. MR **22** #1102.

6. —————————, *Über approximative Nomographie, I, II and III*, Bayer. Akad. Wiss. Math.-Nat. Kl. S.B. (1958), 137-155; *ibid.* (1959), 103-109; *ibid.* (1960), 27-34. MR **22** #1101, **22** #6968, **24** #B1289.

7. M.-B.A. Babaev, *Approximation to functions of several variables by functions of a smaller number of variables*, in "Approximation and Function Spaces," Z. Ciesielski, ed., North-Holland Publ. Co., Amsterdam, 1981, 44-50.

8. —————————, *The approximation of polynomials of two variables by sums of functions of one variable*, (Russian: Azerbaijani summary) Izv. Akad. Nauk Azerbaidzan. SSR Ser. Fiz.-Tehn. Mat. Nauk, no.**2** (1971), 23-29. MR **45** #2377.

9. —————————, *The approximation of polynomials in two variables by functions of the form $\phi(x) + \psi(y)$*, (Russian) Dokl. Akad. Nauk. SSSR **193** (1970), 967-969. Soviet Math. **11** (1970), 1034-1036 (English translation). MR **43** #6634.

10. —————————, *The approximation of functions of several variables by sums of functions of a smaller number of variables in the complex domain*, (Russian: Azerbaijani summary) in "Special Questions on Differential Equations and Function Theory," (Russian) Izdat "Elm," Baku, 1970, 3-44. MR **43** #2398.

11. —————————, *The approximation of a function of several variables by sums of functions of a smaller number of variables in a complex region*, (Russian: Azerbaijani summary) Akad. Nauk Azerbaidzan. SSR Dokl. **23**, no.**2** (1967), 3-7. MR **35** #7049.

12. —————————, *The approximation of a function of several variables by sums of functions of a smaller number of variables in a complex region*, (Russian: Azerbaijani summary) Akad. Nauk Azerbaidzan. SSR Dokl. **23**, no.**1** (1967), 3-8. MR **35** #7048.

13. ——————, *On the best uniform approximation of functions of several variables by sums of functions of one variable in the complex domain*, (Russian) in "Functional Analysis. Certain Problems in the Theory of Differential Equations and Theory of Functions," (Russian) Izdat. Akad. Nauk Azerbaidzan. SSR, Baku, 1967, 24-29. MR **36** #4228.

14. ——————, *On the best approximation in generalized Lebesgue spaces to functions of several variables by sums of functions of a single variable*, (Russian) in "Studies in the Theory of Differential Equations and Theory of Functions," (Russian) Izdat. Akad. Nauk Azerbaidzan. SSR, Baku, 1965, 11-24. MR **34** #8046.

15. ——————, *On the best power approximations of functions of n variables by functions of the type $\phi_1(x_1) + \cdots + \phi_n(x_n)$*, (Russian) in "Studies of Contemporary Problems in the Constructive Theory of Functions," (Proc. Second All-Union Conf., Baku, 1960) (Russian) Izdat. Akad. Nauk Azerbaidzan. SSR, Baku, 1965, 26-33. MR **33** #6226.

16. ——————, *On estimating and determining the value of the best approximation of functions of several variables by sums of functions of one variable in generalized Lebesgue space*, (Russian) in "Problems in Functional Analysis and Applications," (Russian) Izdat. Akad. Nauk Azerbaidzan. SSR, Baku, 1965, 17-22. MR **33** #6224.

17. ——————, *On the best degree of approximation of functions of two variables by functions of the type $\phi(x) + \psi(y)$*, (Russian: Azerbaijani Summary) Izv. Akad. Nauk Azerbaidzan. SSR Ser. Fiz.-Mat. Tehn. Nauk, no.**6** (1962), 25-40. MR **26** #6659.

18. G. Bachman and L. Narici, "Functional Analysis," Academic Press, New York, 1966.

19. J.B. Baillon, R.E. Bruck, and S. Reich, *On the asymptotic behavior of nonexpansive mappings and semigroups in Banach spaces*, Houston J. Math **4** (1978), 1-9.

20. R. Bank, *An automatic scaling procedure for a d'Yakanov-Gunn iteration scheme*, Linear Algebra and its Applications, **28** (1979), 17-33. MR **80h**:65098.

21. ——————, *Marching algorithms for elliptic boundary value problems. II: The variable coefficient case*, SIAM J. on Numerical Analysis **14** (1977), 950-970.

22. A.L. Brown, *Finite rank approximations to integral operators which satisfy certain total positivity conditions*, J. Approximation Theory **34** (1982), 42-90. MR **83f**:41017.

23. R.E. Bruck and S. Reich, *Nonexpansive projections and resolvents of accretive operators in Banach spaces*, Houston Math. J. **3** (1977), 459-469.

24. R.C. Buck, *Approximate functional complexity*, Bull. Amer. Math. Soc. **81** (1975), 1112-1114. MR **52** #8356.

25. ——————, *Approximation properties of vector-valued functions*, Pacific J. Math **53** (1974), 85-94.

26. ————, *Approximation theory and functional equations. II*, J. Approximation Theory **9** (1973), 121-125. MR **51** #13536.

27. ————, *On approximation theory and functional equations*, J. Approximation Theory **5** (1972), 228-237. MR **51** #13535.

28. ————, *On the functional equation $\phi(x) = g(x)\phi(B(x)) + u(x)$*, Proc. Amer. Math. Soc. **31** (1972), 159-161. MR **46** #7746.

29. ————, *Alternation theorems for functions of several variables*, J. Approximation Theory **1** (1968), 325-334. MR **39** #1879.

30. ————, *Functional decomposition and metric entropy*, Expository Report No. 8, Institute for Defense Analysis, Princeton, 1968.

31. ————, *Approximate functional complexity*, in "Approximation Theory II," G.G. Lorentz, C.K. Chui and L.L. Schumaker, eds., Academic Press, New York, 1976, 303-307.

32. ————, *Applications of duality in approximation theory*, in "Approximation of Functions," H.L. Garabedian, ed., Elsevier, Amsterdam, 1965, 27-42.

33. A.S. Cereteli, *Approximation of functions of several variables by functions of the form $\phi_1(x_1) + \phi_2(x_2) + \cdots + \phi_n(x_n)$*, (Russian: Georgian Summary) Thbilis. Sahelmc. Univ. Srom. Mekh.-Math. Mecn. Ser. **129** (1968), 397-409. MR **40** #7679.

34. ————, *Approximation of functions of two variables by functions of the form $\phi(x) + \psi(y)$*, (Russian: Georgian summary) Sakharth. SSR Mecn. Akad. Moambe **44** (1966), 545-547. MR **34** #6401.

35. S. Chevet, *Sur certains produits tensoriels topologiques d'espaces de Banach*, Z. Wahrscheinlichkeitstheorie Verw. Gebiete **11** (1969), 120-138. MR **54** #3436.

36. ————, *Sur certains produits tensoriels topologiques d'espaces de Banach*, C.R. Acad. Sci. Paris Ser. A-B **266** (1968), A413-A415. MR **38** #2574.

37. L. Ciobanu, *The approximation of continuous functions of two variables $f(x, y)$ by polynomials $P(z)$ where $z = xy$*, (Russian: Romanian summary) Bul. Inst. Politehn. Iasi (N.S.) **13** **(17)** (1967), facs. 3-4, 135-138. MR **38** #460.

38. L. Collatz, *Approximation by functions of fewer variables*, "Lecture Notes in Mathematics," W.N. Everitt and B.D. Sleeman, eds., vol. **280**, Springer-Verlag, Berlin, 1972, 16-31. MR **54** #3240.

39. ————, *Zur Tschebyschew-Approximation bei Funktionen mehrerer unabhängiger Veränderlicher*, Proceedings of the Conference on the Constructive Theory of Functions (Approximation Theory) (Budapest, 1969), Akademiei Kiado, Budapest (1972), 89-99. MR **54** #3254.

40. V.A. Daugavet, *Approximation of a function of two variables by the sum of the products of functions of one variable,* (Russian) Metody Vycisl No. **12** (1981), 174-186, 235. MR **82k**:41024.

41. M.M. Day, "Normed Linear Spaces," Second Edition, Springer-Verlag, Berlin, 1973.

42. F.-J. Delvos, *D-variate Boolean interpolation,* J. Approximation Theory **34** (1982), 99-114. MR **83b**:41004.

43. F.-J. Delvos and H. Posdorf, *Generalized Biermann interpolation,* Resultate der Math. **5** (1982), 6-18. MR **84j**:41004.

44. ——————, *Boolesche zweidimensionale Lagrange-Interpolation,* Computing **22** (1979), 311-323. MR **83m**:41002.

45. F.-J. Delvos and W. Schempp, *The method of parametric extension applied to right invertible operators,* Numer. Funct. Analysis and Optimization **6** (1983), 135-148.

46. ——————, *On precision sets of interpolation projectors,* in "Multivariate Approximation II," W. Schempp and K. Zeller, eds., ISNM **61** (1982), 107-124. Birkhäuser-Verlag, Boston, 1982.

47. F. Deutsch, *Von Neumann's alternating method: the rate of convergence,* in "Approximation Theory IV," C.K. Chui, L.L. Schumaker, and J.D. Ward, eds., Academic Press, New York, 1984, 427-434.

48. ——————, *Applications of von Neumann's alternating projections algorithm,* in "Mathematical Methods in Operations Research," P. Kenderov, ed., Bulgarian Academy of Science, Sofia, 1984, 44-51.

49. ——————, *Rate of convergence of the method of alternating projections,* in "Parametric Optimization and Approximation," B. Brosowski and F. Deutsch, eds., ISNM vol. **72**, Birkhäuser, Basel, 1985, 96-107.

50. ——————, *Linear selections for the metric projection,* J. Functional Analysis **49** (1982), 269-292. MR **84a**:41029.

51. ——————, *The alternating method of von Neumann,* in "Multivariate Approximation Theory," W. Schempp and K. Zeller, eds., ISNM vol. **51**, Birkhäuser, Basel, 1979, 83-96.

52. F. Deutsch and P. Kenderov, *Continuous selections and approximate selection for set-valued mappings and applications to metric projections,* SIAM J. Math. Analysis **14** (1983), 185-194. MR **84c**:54026.

53. F. Deutsch, J. Mach, and K. Saatkamp, *Approximation by finite rank operators,* J. Approximation Theory **33** (1981), 199-213. MR **84m**:47058.

54. J. Diestel, "Geometry of Banach Spaces," Lecture Notes in Mathematics, vol. **485**, Springer-Verlag, New York, 1975.

55. J. Diestel and J.J. Uhl, "Vector Measures," The American Mathematical Society, Providence, R.I., 1977.

56. S.P. Diliberto and E.G. Straus, *On the approximation of a function of several variables by the sum of functions of fewer variables,* Pacific J. Math. **1** (1951), 195-210. MR **13**, p. 334.

57. N. Dunford and J.T. Schwartz, "Linear Operators. Part I," Interscience, New York, 1959.

58. N. Dyn, *A straightforward generalization of Diliberto and Straus' algorithm does not work,* J. Approximation Theory **30** (1980), 247-250.

59. L. Flatto, *The approximation of certain functions of several variables by sums of functions of fewer variables,* Amer. Math. Monthly **73** (1966), 131-132. MR **34** #1766.

60. C. Franchetti and E.W. Cheney, *Minimal projections in tensor-product spaces,* J. Approximation Theory **41** (1984), 367-381.

61. ————————, *Best approximation problems for multivariate functions,* Bollettino Unione Mat. Ital. **18-B** (1981), 1003-1015. MR **83a**:41023.

62. ————————, *Simultaneous approximation and restricted Chebyshev centers in function spaces,* in "Approximation Theory and Applications," Z. Ziegler, ed., Academic Press, New York, 1981, 65-88. MR **82f**:41042.

63. C. Franchetti and W.A. Light, *The alternating algorithm in uniformly convex spaces,* J. London Math. Soc. (2) **29** (1984), 545-555. MR **85h**:41064.

64. ————————, *On the von Neumann alternating algorithm in Hilbert space,* J. Math. Analysis and Applications (to appear).

65. D.R. Fulkerson and P. Wolfe, *An algorithm for scaling matrices,* SIAM Rev. **4** (1962), 142-146. MR **25** #1039.

66. D.J.H. Garling, *Absolutely p-summing operators in Hilbert space,* Studia Math. **38** (1970), 319-331.

67. J.E. Gilbert and T.J. Leih, *Factorization, tensor products, and bilinear forms in Banach space theory,* in "Notes in Banach Spaces," H.E. Lacey, ed., The University of Texas Press, Austin, 1980, 182-305.

68. M. Goldstein, W. Haussmann, and K. Jetter, *Best harmonic L^1 approximation to subharmonic functions,* J. London Math. Soc., to appear.

69. M. von Golitschek, *Shortest path algorithms for the approximation by nomographic functions,* in "Approximation Theory and Functional Analysis," P.L. Butzer, R.L. Stens, B. Sz-Nagy, eds., Birkhäuser Verlag, Basel, ISNM vol. **65**, 1984.

70. ————————, *Optimal cycles in doubly weighted graphs and approximation of bivariate functions by univariate ones*, Numerische Math. **39** (1982), 65-84. MR **83m**:41019.

71. ————————, *Remarks on functional representation*, in "Approximation Theory III," E.W. Cheney, ed., Academic Press, New York, 1980, 429-434. MR **82g**:26017.

72. ————————, *Approximating bivariate functions and matrices by nomographic functions*, in "Quantitative Approximation," R. DeVore, K. Scherer, eds., Academic Press, New York, 1980, 143-151. MR **82f**:65144.

73. ————————, *An algorithm for scaling matrices and computing the minimum cycle mean in a digraph*, Numer. Math. **35** (1980), 45-55.

74. ————————, *Approximation of functions of two variables by the sum of two functions of one variable*, in "Numerical Methods of Approximation Theory," L. Collatz, G. Meinardus, H. Werner, eds., Birkhäuser Verlag, ISNM vol. **52**, 1980, 117-124.

75. M. von Golitschek and E.W. Cheney, *Failure of the alternating algorithm for best approximation of multivariate functions*, J. Approximation Theory **38** (1983), 139-143.

76. ————————, *The best approximation of bivariate functions by separable functions*, Contemporary Math. **21** (1983), 125-136.

77. ————————, *On the algorithm of Diliberto and Straus for approximating bivariate functions by univariate ones*, Numer. Funct. Analysis and Optimization **1** (1979), 341-363. MR **80g**:41023.

78. M. Golomb, *Approximation by functions of fewer variables*, in "On Numerical Approximation," R. Langer, ed., University of Wisconsin Press, Madison, Wisconsin, 1959, 275-327. MR **21** #962.

79. H. Gonska and K. Jetter, *Jackson-type theorems on approximation by trigonometric and algebraic pseudopolynomials*, J. Approximation Theory, to appear.

80. W.J. Gordon, *Blending-function methods of bivariate and multivariate interpolation and approximation*, SIAM J. Numerical Analysis **8** (1971), 158-177. MR **43** #8209.

81. ————————, *Distributive lattices and the approximation of multivariate functions*, in "Approximation with Special Emphasis on Spline Functions," I.J. Schoenberg, ed., Academic Press, New York, 1969, 223-277. MR **43** #779.

82. ————————, *Spline-blended surface interpolation through curve networks*, J. Math. Mech. **18** (1968/69), 931-952. MR **39** #7333.

83. W.J. Gordon, J. C. Cavendish, and C.A. Hall, *Ritz-Galerkin approximation in blending function spaces*, Numer. Math. **26** (1976), 155-178. MR **56** #10054.

84. W.J. Gordon and E.W. Cheney, *Bivariate and multivariate interpolation with non-commutative projectors*, in "Linear Spaces and Approximation," P.L. Butzer and B. Sz.-Nagy, eds., Birkhäuser-Verlag, Basel, ISNM vol. **40** (1978), 381-387.

85. W.J. Gordon and C.A. Hall, *Construction of curvilinear co-ordinate systems and application to mesh generation*, Internat. J. Numer. Methods Engrg. **7** (1973), 461-477. MR **56** # 10057.

86. W.F. Gordon and A. Wixom, *Shepard's method of "metric interpolation" to bivariate and multivariate interpolation*, Math. Comp. **32** (1978), 253-264. MR **56** #16230.

87. P.R. Halmos, "Measure Theory," Van Nostrand, Princeton, 1950. Reprinted by Springer-Verlag.

88. E.J. Halton and W.A. Light, *Minimal projections in bivariate function spaces*, J. Approximation Theory (to appear).

89. —————, *Projections on tensor product spaces*, Trans. Amer. Math. Soc. **287** (1985), 161-165.

90. —————, *Minimal projections in L_p-spaces*, Math. Proc. Camb. Phil. Soc. **97** (1985), 127-136.

91. —————, *Existence questions in tensor product spaces*, in "Approximation Theory IV," C.K. Chui, L.L. Schumaker, and J.D. Ward, eds., Academic Press, New York, 1983, 499-504.

92. W. Haussmann, K. Jetter, and B. Steinhaus, *Degree of best approximation by trigonometric blending functions*, Math. Zeit. **189** (1985), 143-150.

93. W. Haussmann and P. Pottinger, *On the construction and convergence of multivariate interpolation operators*, J. Approximation Theory **19** (1977), 205-221.

94. —————, *On multivariate approximation by continuous linear operators*, in "Constructive Theory of Functions of Several Variables," W. Schempp and K. Zeller, eds., Lecture Notes in Mathematics, vol. **571**, Springer-Verlag, Berlin, 1977, 101-108.

95. W. Haussmann and K. Zeller, *Mixed norm multivariate approximation with blending functions*, Proceedings of the Conference on Constructive Theory of Functions held in Varna, Bulgaria, 1984.

96. —————, *Blending interpolation and best L^1-approximation*, Archiv für Math. **40** (1983), 545-552.

97. —————, *Uniqueness and non-uniqueness in bivariate L^1-approximation*, in "Approximation Theory IV," C.K. Chui, L.L. Schumaker, and J.D. Ward, eds., Academic Press, New York, 1983, 509-514.

98. —————, *Multivariate approximation and tensor products*, in "Approximation Theory III," E.W. Cheney, ed., Academic Press, New York, 1980, 495-500.

99. S.J. Havinson, *A Chebyshev theorem for the approximation of a function of two variables by sums* $\phi(x) + \psi(y)$, Izv. Akad. Nauk SSSR Ser. Mat. **33** (1969), 650-666. English translation, Math. USSR, Izv., **3** (1969), 617-632. MR **41**#7351.

100. M.S. Henry and D. Schmidt, *Continuity theorems for the product approximaton operator,* in "Theory of Approximation," A.G. Law and B.N. Sahney, eds., Academic Press, New York, 1976, 24-42.

101. S. Hitotumatu, *On the approximation of a function of two variables by the product,* Information Processing in Japan **8** (1968), 41-43. MR **41** #4081.

102. S.M. Holland, W.A. Light, and L.J. Sulley, *On proximinality in* $L_1(T \times S)$, Proc. Amer. Math. Soc. **86** (1982), 279-282. MR **84j**:41050.

103. R.B. Holmes, "Geometric Functional Analysis and its Applications," Springer-Verlag, Berlin, 1975.

104. —————————, "A Course on Optimization and Best Approximation," Lecture Notes in Mathematics, vol. **257**, Springer-Verlag, Berlin, 1972.

105. I.I. Ibragimov and M.-B.A. Babaev, *Approximation of functions of several variables by the sums of two functions of a smaller number of variables,* in "Approximation Theory" (Proc. Conf. Inst. Math., Adam Mickiewicz Univ., Poznan, 1972) Riedel, Dordrecht, 1975, 97-116. MR **56** #16217.

106. —————————, *Methods for finding functions that deviate the least from functions of several variables,* (Russian) Dokl. Akad. Nauk SSSR **197** (1971), 766-769. Soviet Math. Dokl. **12** (1971), 540-544 (English translation). MR **43** #5219.

107. R.C. James, *Orthogonality and linear functionals in normed linear spaces,* Trans. Amer. Math. Soc. **61** (1947), 265-292.

108. G.J.O. Jameson, "Topology and Normed Spaces," Chapman and Hall, London, 1974.

109. G.J.O. Jameson and A. Pinkus, *Positive and minimal projections in function spaces,* J. Approximation Theory **37** (1983), 182-195.

110. C.T. Kelley, *A note on the approximation of functions of several variables by sums of functions of one variable,* J. Approximation Theory **29** (1980), 305-322.

111. J.L. Kelley, "General Topology," D. van Nostrand, New York, 1955. Reprinted by Springer-Verlag, Berlin.

112. R. Khalil, *Best approximation in* $L^p(I, X)$, Math. Proc. Camb. Phil. Soc. **94** (1983), 277-279.

113. G. Köthe, "Topological Vector Spaces II," Springer-Verlag, New York, 1979.

114. K. Kuratowski and C. Ryll-Nardzewski, *A general theorem on selectors,* Bull. Acad. Polonaise Sciences, Serie des Sciences Math. Astr. Phys. **13** (1965), 397-403. MR **32** #6421.

115. A.J. Lazar, P.D. Morris, and D.E. Wulbert, *Continuous selections for metric projections*, J. Functional Analysis **3** (1969), 193-216.

116. W.A. Light, *A note on proximinality in $C(S \times T)$ with the L_1-norm*, Proc. Edinburgh Math. Soc., to appear.

117. ———————, *The Diliberto-Straus algorithm in $L_1(X \times Y)$*, J. Approximation Theory **38** (1983), 1-8. MR **84h**:41048.

118. ———————, *Convergence of the Diliberto-Straus algorithm in $L_1(X \times Y)$*, Numer. Funct. Analysis and Optimization **3** (1981), 137-146. MR **84a**:41042.

119. ———————, *Central proximity maps*, in "Approximation Theory III," E.W. Cheney, ed., Academic Press, New York, 1980, 589-594. MR **82g**:41033.

120. W.A. Light and E.W. Cheney, *The characterization of best approximations in tensor-product spaces,* Analysis **4** (1984), 1-26.

121. ———————, *Some best approximation theorems in tensor-product spaces*, Math. Proc. Camb. Phil. So. **89** (1981), 385-390. MR **82d**:41045.

122. ———————, *On the approximation of a bivariate function by the sum of univariate functions*, J. Approximation Theory **29** (1980), 305-322. MR **82d**:41023.

123. W.A. Light and S.M. Holland *The L_1-version of the Diliberto-Straus algorithm in $C(S \times T)$*, Proc. Edinburgh Math. Soc. **27** (1984), 31-45.

124. W.A. Light, J.H. McCabe, G.M. Phillips, and E.W. Cheney, *The approximation of bivariate functions by sums of univariate ones using the L_1-metric*, Proc. Edinburgh Math. Soc. **25** (1982), 173-181. MR **84k**:41036.

125. P.-K. Lin, *Remarks on linear selections for the metric projection*, J. Approximation Theory **43** (1985), 64-74.

126. J. Lindenstrauss and L. Tzafriri, "Classical Banach Spaces," Lecture Notes in Mathematics, vol. **338**, Springer-Verlag, Berlin, 1973.

127. G.G. Lorentz, *The 13th problem of Hilbert,* in "Mathematical Developments Arising From Hilbert Problems," F.E. Browder, ed., Proceedings of Symposia in Pure Mathematics, Amer. Math. Soc. **28** (1976), 419-430.

128. V.N. Malozemov, *The approximation of continuous functions of two variables by algebraic polynomials*, (Russian: English summary) Vestnik Leningard. Univ. **23**, no.19 (1968), 73-87. MR **39** #682.

129. C.A. Micchelli and G. Wahba, *Design problems for optimal surface interpolation*, in "Approximation Theory and Applications," Z. Ziegler, ed., Academic Press, New York, 1981, 329-348.

130. E. Michael, *Continuous selections*, Ann. Math. **63** (1956), 361-382.

131. ———————, *Selected selection theorems*, Amer. Math. Monthly **63** (1956), 233-238.

132. V.M. Mordasev, *The best approximation of a function of several variables by a sum of functions of fewer variables*, (Russian) Mat. Zametki **5** (1969), 217-226. MR **39** #4571.

133. ———————, *Approximation of a function of several variables by a sum of functions of a smaller number of variables*, (Russian) Dokl. Akad. Nauk SSSR **183** (1968), 778-779. Soviet Math. Dokl. **9** (1968), 1462-1463 (English translation). MR **38** #3666.

134. I.P. Natanson and S.A. Agahanov, *The best approximations of continuous functions of two variables*, (Russian) Leningrad. Meh. Inst. Sb. Naucn. Trudov **No. 50** (1965), 15-18. MR **39** #1874.

135. J. von Neumann, "Functional Operators, Vol. II," Annals of Mathematics Studies **22**, Princeton University Press, 1950.

136. A.G.O. O'Farrell, *Five generalizations of the Weierstrass approximation theorem*, Proc. Royal Irish Acad. Sect. A **81** (1981), 65-69. MR **82k**:44004.

137. J.P. Ofman, *Best approximation of functions of two variables by functions of the form $\phi(x) + \psi(y)$*, Amer. Math. Soc. Translations **44** (1965), 12-29. MR **23** #A2684.

138. K.R. Parthasarathy and K. Schmidt, "Positive Definite Kernels, Continuous Tensor Products, and Central Limit Theorems of Probability Theory," Lecture Notes in Mathematics, vol. **272**, Springer-Verlag, Berlin, 1972.

139. T. Parthasarathy, "Selection Theorems and Their Applications," Lecture Notes in Mathematics, vol. **263**, Springer-Verlag, New York, 1972.

140. A. Persson, *On some properties of p-nuclear and p-integral operators*, Studia Math. **33** (1969), 213-222.

141. A. Pietsch, "Operator Ideals," North-Holland Publ. Co., Amsterdam, 1980.

142. A. Pinkus, "The Theory of n-Widths," Ergebnisse Series, Springer-Verlag, New York, 1985.

143. M.J.D. Powell, *Numerical methods for fitting functions of two variables*, in "The State of the Art in Numerical Analysis," Academic Press, New York, 1977, 563-604.

144. H. Reiter, *Contributions to harmonic analysis, VI*, Ann. of Math. **77** (1963), 552-562.

145. J.R. Respess and E.W. Cheney, *On Lipschitzian proximity maps*, in "Nonlinear Analysis and Applications," S.P. Singh and J.H. Burry, eds., Lecture Notes in Pure and Applied Mathematics, vol. **80**, Marcel Dekker, New York, 1982, 73-85. MR **84g**:41033.

146. ———————, *Best approximation problems in tensor-product spaces*, Pacific J. Math. **102** (1982), 437-446. MR **84b**:41030.

147. T.J. Rivlin and R.J. Sibner, *The degree of approximation of certain functions of two variables by the sum of functions of one variable*, Amer. Math. Monthly **72** (1965), 1101-1103. MR **32** #4450.

148. W. Rudin, "Real and Complex Analysis," Second Edition, McGraw-Hill, New York, 1974.

149. —————, "Functional Analysis," McGraw-Hill, New York, 1973.

150. —————, *Projections on invariant subspaces*, Proc. Amer. Math. Soc. **13** (1962), 429-432.

151. P. Saphar, *Produits tensoriels d'espaces de Banach et classes d'applications lineaires*, Studia Math. **38** (1970), 71-100. MR **43** #878.

152. —————, *Produits tensoriels topologiques et classes d'applications lineaires*, C.R. Acad. Sci. Paris Ser. A-B **266** (1968), A526-A528. MR **38** #2575a.

153. H. Schaefer, "Topological Vector Spaces," Macmillan, New York, 1966.

154. R. Schatten, "A Theory of Cross-Spaces," Princeton University Press, 1950. Reprinted by Kraus Reprint Corporation, Milwood, New York, 1965. MR **12**, p. 186.

155. K. Scherer and K. Zeller, *Bivariate polynomial approximation*, in "Approximation and Function Spaces," Z. Ciesielski, ed., North-Holland Publ. Co., Amsterdam, 1981, 621-628.

156. —————, *Gestufte Approximation in zwei Variablen*, in "Numerische Methoden der Approximationstheorie," ISNM vol. **52**, Birkhäuser-Verlag, Basel, 1980.

157. L.L. Schumaker, *Fitting surfaces to scattered data*, in "Approximation Theory II," G.G. Lorentz, C.K. Chui, and L.L. Schumaker, eds., Academic Press, New York, 1967, 203-268.

158. Z. Semadeni, "Banach Spaces of Continuous Functions," Polish Scientific Publishers, Warsaw, 1971.

159. I. Singer, "The Theory of Best Approximation and Functional Analysis," Society for Industrial and Appl. Math., Philadelphia, 1974.

160. —————, "Best Approximation in Normed Linear Spaces by Elements of Linear Subspaces," Springer-Verlag, New York, 1970.

161. B. Steinhaus, L_1-*approximation with blending functions*, preprint, 1984.

162. F. Sullivan, *A generalization of best approximation operators*, Ann. Mat. Pura Appl. **107** (1975), 245-261. MR **53** #3564.

163. V.N. Temlyakov, *On best approximation of functions of two variables*, Anal. Math. **2** (1976), 231-234.

164. —————, *Best approximations of functions of two variables*, (Russian) Dokl. Akad. Nauk SSSR **223** (1975), 1079-1082.

165. F. Treves, "Topological Vector Spaces, Distributions and Kernels," Academic Press, New York, 1967.

166. D.H. Wagner, *Survey of measurable selection theorems: an update,* in "Measure Theory Oberwolfach 1979," D. Kölzow, ed., Lecture Notes in Mathematics, vol. **794,** Springer-Verlag, Berlin, 1980, 176-219.

167. S.E. Weinstein, *Product approximations of functions of several variables,* SIAM J. Numer. Analysis **8** (1971), 178-189.

INDEX OF NOTATION

INDEX

Norm, 3

    p-nuclear, 27

    α, 4

    α_p, 27

    β, 22

    γ, 6

    λ, 3

Null set, 113

Orthogonal projection, 59

P-nuclear norm, 27

Path, 69

Projections

    minimal, 91, 103

    orthogonal, 25

Projection constant, 91

Property B, 97

Proximinal, 35

Proximity map, 35, 48

Reasonable norm, 4

Schatten, R., 138

Section of a function, 11, 60

Selection theorem

    Michael's 133, 37

    Kuratowski-Ryll-Nardzewski, 133

Simple function, 11, 113

Sitting-Duck theorem, 44, 134

Smooth

    point, 48

    space, 48

    subspace, 54

Strongly measurable, 113

Tensor product, 1

    algebraic, 2

    completed, 9

    of Hilbert spaces, 20

    of operators, 19

Uniform crossnorm, 8

Uniformly convex, 50

Von Neuman algorithm, 48, 60

Weakly measurable, 38